THERMAL POOLS
IN ICELAND

Jón G. Snæland and
Þóra Sigurbjörnsdóttir

THERMAL POOLS
IN ICELAND

SKRUDDA

2010

Thermal pools in Iceland
© Jón G. Snæland and Þóra Sigurbjörnsdóttir 2010

Layout and design: Skrudda
English translation: Wincie Jóhannsdóttir
Maps: Sigurgeir Skúlason
Type: Adobe Garamond Pro 10,5/13,5
Printed in Ísafoldarprentsmiðja, Iceland
Published by Skrudda
skrudda@skrudda.is

ISBN: 978–9979–655–66–4

Contents

Maps

Introduction

This book is written with two aims in mind: on the one hand, to introduce visitors to some especially interesting options on their travels and on the other, to encourage municipalities and enthusiasts to respect and attend to our thermal pools. Hopefully it will inspire the public-spirited to take neglected and dried up pools under their wing, much as they have often done with old highland huts or lodges and various lowland antiquities.

The book covers thermal pools, whether natural or man-made, that are well worth visiting. Some are in built-up areas, others in the wilderness. This is not an exhaustive survey of Iceland's natural thermal pools, since there are around 700 geothermal areas, each of which may contain any number of thermal springs. The water in them ranges from boiling to near body temperature to tepid. Only a few decades ago, there were many more natural pools suitable for bathing, but in recent years various projects, such as drilling for hot water to heat houses, have reduced their numbers. Others have given way to new,

Helena Auðbjörg Snæland by the pool at Lindir which disappeared under the Hálslón reservoir. Photo: Jón G. Snæland

The ruins of the Leirárlaug pool, with Hafnarfjall in the background. Photo: Jón G. Snæland

concrete pools. Travellers are keen to visit these wild pools, both summer and winter. It can often be difficult to find them, since they tend to blend fairly quietly into the landscape, and are often at a great distance from anything man-made. For this reason the GPS position of each pool is to be found in this book. Low-temperature geothermal

Looking down towards the farm Presthvammur. The now-empty Birningsstaðalaug can be seen across the river Laxá.Photo: Þóra Sigurbjörnsdóttir

areas are not equally common in all parts of the country; whereas there are few pools in the Eastern fjords there is abundance in the Western fjords, where there are a great many small thermal pools, especially in Ísafjarðardjúp. The man-made pools were built at different times, are variously equipped, and often have a rich history. The reasons they were built vary, a number deriving from the passion youth-associations in the early 20th century had for improving Icelanders' swimming abilities.

We wish to thank the many people who have assisted us in compiling this book, providing photographs and other help. In particular we thank Kjartan Gunnsteinsson for allowing us access to his photo collection and Jón Þorsteinsson for reading over the manuscript.

The Highlands

Hveravellir

GPS N64 45.974-W19 33.228 WGS 84

627 meters asl

The pool at Hveravellir is in a warm stream below the Old Lodge. It is about 6 x 3.5 m, built up with concrete and rocks with flat rock at the bottom. At its deepest it is 1.35 m. Some of the water is piped from the stream at a rate of 0.9 l/sec. There is no sediment to cloud the pool, but it is coloured by the minerals dissolved in the hot water. The temperature of the pool is between 18.6 and 39.3°C, and the water in the intake pipe is 80°C. There is room for several dozen bathers but there are no changing facilities by the pool and it is forbidden to go into the Old Lodge to change. Bathers change outside the Old Lodge or in nearby restrooms. Those who change by the Old Lodge can leave their clothes on the wooden deck. One can't help wondering at these arrangements and why changing rooms by the pool have not been built since Hveravellir often has several hundred visitors a day.

A busy day at the nearly 60-year-old thermal pool at Hveravellir. Photo: Jón G. Snæland

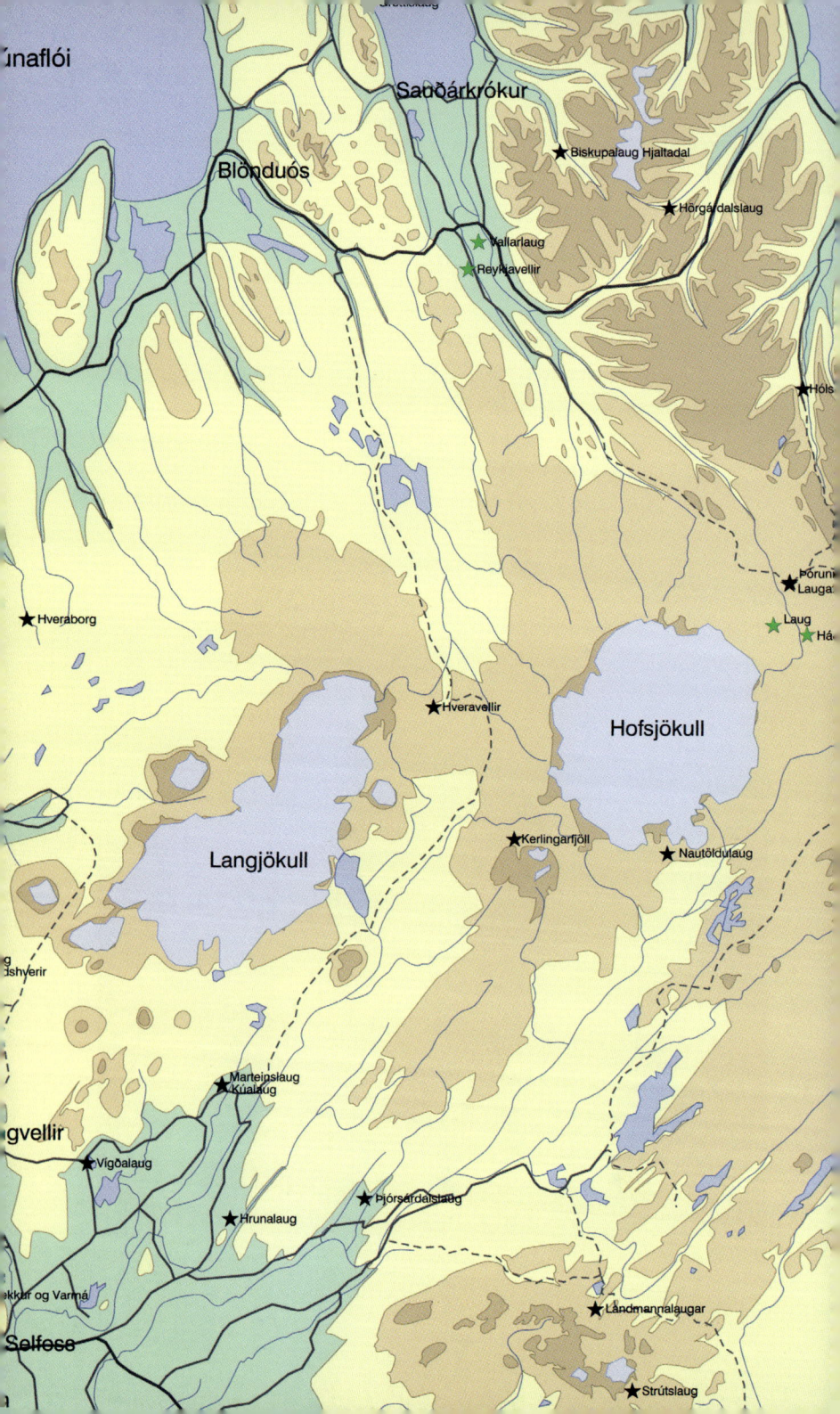

Húnaflói

Sauðárkrókur

Blönduós

★ Biskupalaug Hjaltadal

★ Hörgárdalslaug

☆ Vallarlaug

☆ Reykjavellir

★ Hóls

★ Þórun
Laug

★ Hveraborg

☆ Laug
★ Há

★ Hveravellir

Hofsjökull

Langjökull

★ Kerlingarfjöll

★ Nautöldulaug

shverir

★ Marteinslaug
★ Kúalaug

gvellir

★ Vigðalaug

★ Þjórsárdalslaug

★ Hrunalaug

kkur og Varmá

Selfoss

★ Landmannalaugar

★ Strútslaug

aflastaðir

Presthvammslaug
Birningsstaðalaug

Hjalli

Stóragjá
Vogagjá
Jarðböðin
Grjótagjá

vri

Reykjalaug

Viti

Laugavallalaug
Laugarhús

Laufrandarlaug

Lindúr

Laugarfell

Hítulaug

Hveragil

Snapadalur

Vatnajökull

Grímsfjall

Horna

Guðlaug

The Highlands

Pools are marked with a star.
Pools marked with a green star
are not specifically covered.

Skeiðarársandur

Öskurhólshver (Bellowing Spring). Photo: Þóra Sigurbjörnsdóttir

According to Páll Arason, driver and mountain-travel entrepreneur, it was Gunnar Stefánsson, father of Árni Gunnarsson MP, who suggested around 1950 that the thermal stream should be dammed to make a bathing pool. It appears to have been redone in its present form in 1958 and in 1960 the pool and the entire hot spring area were designated a conservation area.

The hot spring area

Hveravellir is what is known as a high-temperature area, that is, a geothermal area where water reaches a tempereature of 150°C at a depth of 1000 m. It stands about 630 m asl in a dale between the Kjalarhraun lava field and Breiðimelur. The water table at Hveravellir being rather high, most of the springs produce water (not steam). The oldest known description of the area is from the 1752 scientific expedition of Eggert Ólafsson and Bjarni Pálsson. They describe the thermal springs in the area, especially one that Eggert named Öskurhólshver (Bellowing Spring) because of the booming and whistling sounds it produced at that time, though it does so no longer.

The pool has been a conservation area since 1960. Photo: Þóra Sigurbjörns-dóttir

The oldest known record of people travelling by the Kjalvegur route refers to reconnaissance trips around the year 900. At that time Hvera-vellir was called Reykjavellir. By the 12th century Kjalvegur was a busy route, and journeys along it are often mentioned in Sturlunga Saga; it looks as if there was already a rest house there at that time, which might have stood where the present day turf house is. Grett-ir Ásmundarson is supposed to have lived for a while about two km south of Rjúpnafell at Hveravellir, in a cave in a high lava hill. There are many cairns on the hill, and the cave, open at both ends, is known as Grettishellir Cave. Just to the west of the hot spring area at Hvera-vellir are the ruins of a hut where the fugitive outlaws Eyvindur Jóns-son (Fjalla-Eyvindur) and Halla Jónsdóttir are thought to have dwelt, probably early in their wanderings. The hut is in a crevice in the lava, with rocks piled along the edges and forming end walls. The ruins are in two parts, the larger 4.2 x 1.2 m, the smaller 1 x 2 m. Very near the ruins is the Eyvindarhver hot spring where Eyvindur and Halla are said to have boiled their food. Legend has it that Arnes Pálsson and

The warm stream. Beyond it, the hot spring area. Photo: Þóra Sigurbjörns-dóttir

Abraham were with them at Hveravellir, but this conflicts with known dates in the lives of these unfortunates.

It is not clear how long Fjalla-Evindur and Halla dwelt at Hveravell-ir. If it is correct that they went there as soon as they fled, it was prob-ably 1761. The people of Skagafjörður drove them away and they are thought to have gone onto Arnavatnsheiði Heath. On the other hand, it is on record that their shack was found under the Arnarfell mountain in 1762, so their stay there was probably a matter of months, not years. There was another fine outlaw who dwelt at Hveravellir called Magnús Soul-in-peril who is the source of a well-known saying. Magnús had got himself in some kind of trouble he wished to avoid and he fled to Hveravellir. On the way he grabbed a lamb for nourishment during his outlawry. When the time came to slaughter the lamb it bleated so piteously that Magnús hesitated and could hardly bring himself to kill it. But hunger called for action and he famously declared, "No mercy from Magnús" and cut its throat. When it was time to cook dinner he threw the carcass in a hot spring, but it sank to the bottom and nothing floated back up except the lungs. They sustained him for the first week. The second week he was sustained by his own saliva. For the third week, his only sustenance was the grace of God. Claiming that was the most difficult week of all, he gave up on outlawry and went home to meet his fate.

Just to the south of the Touring Association lodge at Hveravellir is the old turf lodge. It was built in 1922 under the auspices of the Ministry of Transport, and stands atop ruins that are thought to be from the time of the Sturlungs. It now belongs to the Svínavatn Rural District but is in the custody of Minjavernd, a heritage preservation company. Having been rebuilt in 1994 this lodge is in very good shape.

In 1938 the Icelandic Touring Association put up a new building a little further to the north, which was added to and improved in 1968.

Near that lodge, on Breiðamelur, the Icelandic Met Office built a weather station, after which they kept someone on duty at Hveravellir all year round. There is also petrol for sale there. The Icelandic Touring Association built yet another lodge in 1980, which is simply called The New Lodge (nýi skálinn), the older one being called The Old Lodge (gamli skálinn). Later the Touring Association put up a big building for restrooms and other services of various kinds. Recently the Svínavatn Rural District won the right to run the services at Hveravellir.

The Borehole in the Kerlingarfjöll Mountains

GPS N64 40.415-W19 17.605

760 meters asl

The most commonly used approach to this pool is from the Kjalvegur about 10 km along a road leading to the Kerlingarfjöll lodge. It is passable by all vehicles in the summer, though there are some streams to be forded, as well as the River Ásgarðsá. There are some sizeable culverts in the river which are always kept in shape over the summer, though soil is washed away from them over the winter and during the spring thaw. Continuing from the lodge you drive in the Ásgarðsá gorge, which means fording the river a number of times. You can also walk from the lodge to the pool in about 15-20 minutes. Another driving route is along the Hrunamannaafréttur and Leppstungur, but this is passable only by jeeps. The same is true of the Illahraun route where you drive west of the River Þjórsá, into Þjórsárver and past Setrið, a

Once the borehole had been drilled they constructed a pool around it, using mostly canvas and stones. Photo: Jón G. Snæland

Here the pool has been built around the borehole and Óskar Erlingsson checks the temperature. Photo: Guðbjörg Sigrún Gunnarsdóttir

lodge belonging to the 4 x 4 Travel Club by the Hofsjökull glacier, and then over the Illahraun lava field to Ásgarður.

The pool stems from when they drilled for hot water for the Kerlingarfjöll centre. Unfortunately the water wasn't hot enough to heat the buildings, but the borehole provides 20 l/sec of warm water that is just right for bathing in the summer. To begin with they constucted a rough pool around the borehole using canvas and stone walls. Later they made a pool lined with upright stone slabs supported by rocks on the outside. It is 2 x 4 m, 0.65 m at its deepest, comfortably taking 10-15 bathers. There are no changing facilities but stone slabs have been laid nearby to leave clothes on.

This pool is exceptional for various reasons. It is situated in a deep gorge with colourful rhyolite cliffs, and sitting in a pool where water is powerfully gushing up in the middle is a most unusual experience. There is a cluster of around 20 buildings of various kinds in the Ker-

A shelter for farmers fetching their sheep from the highlands was built in Kerlingarfjöll in 1890. It was situated at the gable end of the rest-house built by the Touring Association in 1936. At that time it could sleep 16, but it was enlarged in 1959 and now sleeps 40-50, 26 in bunks. Photo: Bjarni Kristinsson

lingarfjöll mountains where Fannborg Ltd. provides all kinds of services over the summer. You can get accommodation there, and in the main lodge they sell food and petrol.

Laugafell

GPS N65 01.648-W18 19.920 WGS 84
744 meters asl

Below the north side of the mountain Laugafell, northeast of Hofs-jökull glacier, are a number of lodges which together are called Lauga-fellsskálarnir. The westernmost, named Hjörvarsskáli, is privately owned by snowmobile enthusiasts from Eyjafjörður. Further up is a two-storey lodge with dressing rooms and toilets on the ground floor and sleeping accommodation upstairs. These services are considered among the best in the highlands, perhaps because the building is heated all year, using geothermal heat. The lodge belongs to the Touring Club of Akureyri, as does the main lodge, Laugafell, a bit further north. The southernmost lodge is a small hut for the warden who is at Laugafell over the summer. There is also a camp site at Laugafell.

A peaceful morning at Laugafell. Photo: Jón G. Snæland

Winter reigns at Laugafell. Photo: Jón E. Halldórsson

Between the lodges, below a lovely ridge, is a large pool, built of turf and stones. It was built in 1976 by members of the Touring Club of Akureyri and has been improved twice since then, most recently in 2000, by snowmobile enthusiasts from Eyjafjörður who also see to the maintenance of the pool, which they always clean out on their first working trip each summer. This is most necessary, since over the winter algae fills the pool, making it so slippery towards the end of winter that it is almost impossible to keep your footing. The pool is 16 x 7 m, so it can take dozens of bathers The temperature is 33–38°C, the depth from 50 to 150 cm and the flow 5 l/sec. There is another thermal pool at Laugafell, Þórunnarlaug (see p. 27).

In the summer there are a number of approaches to Laugafell. You can take road no. F752, Skagafjarðarleið (Sprengisandsleið) from Skagafjörður, but if the glacial river Hnjúkskvísl is in spate it can become impassable to all but altered jeeps. From Eyjafjörður you can take road no. F821 which is passable in summer by small jeeps and cars with relatively higher clearance. From the east, road no. F881,

Dragaleið, connects Laugafell with Sprengisandsleið. This road is passable by all vehicles. Finally, you could take Forsetavegur, road no. F752, from the south, which also connects with Sprengisandsleið. On Forsetavegur you have to ford one of the source branches of the river Þjórsá, Bergvatnskvísl. On the whole, the water is low in this river and easily forded by smaller jeeps, though a long wet spell can make it an obstacle. Travellers who plan to stay several days at Laugafell thus have a number of options for day-trips.

There are several famous old routes near the lodges. If you are driving a jeep or are a good walker you should maybe start by going up the mountain Laugafell, only a 2 km walk and easily passable for all jeeps. The track ends by a cairn at the very top, 904 m asl. From there you can see east to Trölladyngja and Vatnajökull, and Sprengisandur lies at your feet along with Hofsjökull to the south. You can take the renowned route Eyfirðingavegur: drive along Skagafjarðarleið and cross the bridge over the river Austari-Jökulsá. A little north of the bridge there is a track to the left of the main track, south of the grasslands Orravatnsrústir.

This will bring you to the so-called Strompleið (Smokestack Road), which takes its name from an old herring factory smokestack used as a culvert, which you come to quite soon. Though the smokestack has been replaced with a modern culvert it is still the most likely obstacle on this route since the fill often gets washed away. Once you've crossed the culvert the track is clear all the way across to Eyfirðingavegur at Rauðhólar by Ásbjarnarvötn. The way continues clear west as far as the building Ingólfsskáli on the easternmost branch of the river Vestari-Jökulsá, called Austari-Krókkvísl. This is the end of the road for all but those who are highly experienced at fording Icelandic rivers and driving an extremely altered jeep, since there are many dangerous glacial rivers to be crossed if you carry on to the west. Those who wish to drive on, however, can continue south along the river to Hofsjökull via a track laid for scientists who were recording precipitation on the glacier.

From Laugafell you can also take a short trip down to the Gráni hut by driving down along the river Laugakvísl on the east, then along Austari-Jökulsá and crossing Geldingaá. On the river bank are two

Hjörvarsskáli at Laugafell belongs to snowmobile enthusiasts from Eyjafjörð-ur

huts, the modern Sesseljubúð and the turf and stone built Gráni, built in 1920 and now used to house animals. It has an unusual history. The housewife at the farm Jökull in Eyjafjörður had got tired of farmers using her favourite riding horse every autumn when they went to fetch their sheep from the highlands in cold, wet, uncomfortable conditions. For her horse's comfort she had them build a rest house, superior to most in the north. She chose the best place and named the hut for the horse, Gráni. From there you can circle back to Laugafell. On the way a track can be seen to the left of the road. It leads over to Vatnahjallaleið (see Hólsgerðislaug p. 149), and can be followed south to join the main route out of Eyjafjörður.

Þórunnarlaug

GPS N65 01.822-W18 19.770 (imprecise)
750 meters asl

Beyond the road at the northern corner of the parking lot by the main lodge at Laugafell you will find a path leading to a little gorge with a stream in the bottom. Follow the stream up along the gorge for 150-200 m and you will find a little thermal pool just big enough for one person, who may only be of medium stature to fit comfortably into the pool. Some say that Þórunn from Grund had the pool hewn out when she lived at Laugafell, others disagree and Þorvaldur Thoroddsen, for instance, deemed it a natural pool.

The story is that Þórunn from Grund settled at Laugafell with her household at the time of the plague. That story loses some credibility from the fact that the plague arrived in Iceland (Hvalfjörður) in 1402, but Þórunn wasn't born til a century later. A different story claims

The Þórunnarlaug thermal pool was considered to have curative powers. Photo: Jón G. Snæland

The pool of Þórunn the rich. Photo: Jón G. Snæland

this housewife was Þórunn the rich from Möðruvellir. Guðbrandur Þorkell Guðbrandsson in Sauðárkrókur says. "It is common in stories for people to confuse the two big plagues of the fifteenth century, one of which is said to be from 1420, while the other came in the early 1490s, as far as we know. The problem is that sources from that period are few, partly because they were lost in fires. The fact that there was nearly a century between plagues might explain the discrepancy regarding Þórunn's time at Laugafell."

Ruins of buildings have been found on the banks of Laugakvísl. According to old oral accounts Þórunn the Rich had her mountain dairy (milking sheep) there. Elderly residents of Eyjafjörður name these ruins for Þórunn, and some signs of habitation have been found there.

Nautöldulaug (Ólafslaug)

GPS N64 38.765-W18 46.200
599 meters asl

The pool is in the Þjórsárver moorland (formerly called Fuglafit) below the Nauthagajökull glacial tongue on the south of the glacier Hofsjökull. The pool, beside a low gravel swell a few hundred meters from the end of the tongue, was made by some 4x4 drivers several years ago, who piled sandbags round a small hot spring. The spring melt of 2003 washed them away but members of the Iceland Glaciological Society came the following year to measure the glacial tongues of Hofsjökull and they positioned a fisherman's tub at the spring, making it usable again. The water, running into the tub at 0.2 l/sec, is about 47°C, which is rather too hot for bathing, so people usually shovel snow in to cool it when they visit in winter.

This pool can only be reached by altered jeeps, and even then only in the winter or late autumn when the ground is frozen but free of

Nautöldulaug before Mother Nature swept it away in a flood. Photo: Kjartan Gunnsteinsson

29

Travellers by the old lodge at Nautalda. Photo: Jón G. Snæland

snow, because then you can follow the tracks. It is also possible to walk there in the summer, but the route is precarious because of several fast-flowing glacial rivers, and actually passable only by those with a great deal of highland desert experience.

There are more thermal pools in the Þjórsárver moorland. About an hour's walk from the Nautölduskáli hut in along the west side of Mt. Ólafsfell will bring you to more pools that some people consider cleaner.

There are four ways to approach the Þjórsárver moorland, which all meet up by the river Hnífá on Fjórðungssandur. The first leads from the Kerlingarfjöll Mountains across the Illahraun lava field to Setrið, the Hofsjökull lodge of the 4x4 Travel Club. It continues from there south across Fjórðungssandur to the Þjórsárver turnoff (GPS N64 34.244-W18 55.522). The second is called Klakksleið, from south of the Kerlingafjöll mountains, and it joins the first just south of Setrið. Thirdly you can take Gljúfurleitsleið along the west of the Þjórsá River onto Fjórðungssandur, and finally you can drive along the east of Þjórsá on Kvíslaveituvegur, fording the river at Sóleyjarhöfðavað and going

Hlynur Snæland Lárusson and Bergþór Júlíusson by the new fisherman's tub which relplaced the pool that was swept away in a spring spate or a small glacial burst from Nauthagajökull. Photo: Jón G. Snæland

along Tjarnarver to Þingmannaleið (GPS N64 31.367-W18 53.019), which you then take to the north until it joins the other routes, just north of the river Hnífá. From that crossroads you drive down to the ford in Hnífá. The water in the ford is usually clear, except when the glacial river Blautakvísl branches into Hnífá. The ford is a good one, the river bed pretty solid, but it can be a bit of a scramble up the eastern bank, where the tire tracks are pretty deep. From there you go along the crest of Steingrímsalda to Blautakvísl. From the crest there is a magnificent view over Hofsjökull and Þjórsárver, so it is well worth going that far even if you go no further. At Blautakvísl, after fording a fairly fast-flowing branch of the river, you drive along the river and then east along a branch that runs towards Nautalda. Quicksands are common in this area, so the river beds are often more dependable than the banks.

Just to the south of Nautalda you cross the last branch of Blautakvísl which will bring you to a well-defined track along the south of the rise. To the south-east of the rise is the old National Energy Authority lodge, or rather the middle of the lodge, because there used to be two lodges with a corridor between, but now only the corridor is left, and is used as a mountain hut in its own right. Travellers should take the trouble to walk up onto the Nautalda rise to see the old goose traps – there are quite a few such traps in Þjórsárver, mostly on the crests of swells. The view from there is marvellous; to the north you can often see ice caves in the glacier and a lagoon from which Miklakvísl flows out of Hofsjökull. Since it seldom freezes, even in winter, you can usually see the river running along the west of the mountain Ólafsfell, then meandering past the Nautalda hut and over the moorlands to the Þjórsá River. Miklakvísl can be a serious obstacle in the winter, often with very high ice floes along the banks. Sometimes one can find ice thick enough to hold, even a considerable way above the hut. Once across the river it is fairly easy to reach the thermal pool in the winter, and in the autumn the river is mostly easy to ford and the tracks and

Skúli H. Skúlason where the glacial stream Ólafskvísl comes out of Nauthagajökull not far from Nautöldulaug. Photo: Jón G. Snæland

At the edge of Nauthagajökull. Photo: Magni Rúnar Þorvaldsson

fords well-defined. Having reached Nautöldulaug, travellers should make their way to the glacier tongue, which can be impressive, often with a large ice cave where the minor glacial stream Ólafskvísl comes out of Nauthagajökull. The glacier tongue is, however, very changeable, and the author has been there both when he could virtually have driven onto it in a family car and when he was faced with a sheer cliff of ice. To the east of the thermal pool, across the Múlakvísl gravel banks and along Múlajökull glacier, there is a track which is mostly visible alongside the glacier but tends to disappear on the gravel banks. To the east of Múlajökull you cross Arnarfellskvísl, which can be deep and fast-flowing with steep gravel banks at the ford. From there the mountain Arnarfell the Great rises before you. Its southern slope, called Arnarfellsbrekka, is covered with vegetation, despite the fact that the foot of the mountain is at 620 m asl.

The outlaw Fjalla-Eyvindur is said to have lived here, west of Arnarfell. People say that 15 men came after Eyvindur, Halla and their child, and Arnes and Abraham. Eyvindur fled with his folk onto a glacier, probably Múlajökull, and eventually the sheriff's men found an outlaw's hut and tracks leading onto Hofsjökull. They destroyed the hut with all its contents.

Landmannalaugar

GPS N63 59.516-W19 03.713

593 meters asl

The natural thermal pool at Landmannalaugar in the Fjallabak region gets well over 100.000 visitors a year, and is undoubtedly the most popular thermal pool among both native and foreign travellers in Iceland. From the north and west it is approached along road no. F208, Fjallabaksleið nyrðri (Sigölduleið), which connects the highland hydroelectric station at Hrauneyjar to Landmannalaugar, or along road no. F225, Landmannaleið (Dómadalsleið). From the east Landmannaleið (Dómadalsleið) is reached from Skaftafellssýsla County. All these roads are passable by small jeeps and they meet at Jökulgilskvísl. There you take road no. F224, which fords the small streams Námskvísl and then Laugakvísl, which are on the whole no hindrance, though Námskvísl can be an obstacle for smaller jeeps when it rains heavily. However, it is not necessary to ford these streams, since vehicles can

The bathing pool at Landmannalaugar at the edge of the Laugahraun lava field. Photo: Jón G. Snæland

Bathers changing on the deck by the pool. The more modest can also use the bathhouse by the lodge. Photo: Jón G. Snæland

be left in a parking lot at Námskvísl from which there is a footbridge to the pool, which is at end of the Laugahraun lava field.

The water in the pool comes from dammed streams, both hot and cold, from under the lava. The pool is very large, with room for dozens of bathers at once, and the temperature is 34°-41°C.

Snails are found in the pool, and wading birds sometimes come to Landmannalaugar, so there is some danger of being troubled by flukes, parasites that bore their way into the skin.

There are no changing facilities, but bathers can keep their clothes and other possessions on a large timber deck right beside the pool. There is a wide wooden path from the pool to the Icelandic Touring Association's lodge where the more modest can change in a large bath house with showers for all. There are also first class grilling facilities there.

There are a number of buildings at Landmannalaugar. The first was a turf hut, built in 1850. Another was built in 1905, right at the edge of the lava – it collapsed under the weight of snow, was rebuilt in 1907 but again collapsed under snow. Later they built a small house with

The old sheep herders' hut on the edge of the Laugahraun lava field. Photo: Jón G. Snæland

a sleeping platform and facilites for horses, which can be seen above the main lodge. In 1951-52 the Touring Association built a lodge, a concrete floored corrugated iron quonset hut with room for 20-30. It is now used as a stable and is owned by the Holt and Landmanna local councils. The present Touring Association lodge was built in 1969,

In 1941 a young taxi driver, Guðmundur Sveinsson, having heard of the natural beauty of the Landmannalaugar area, decided to invite the famous painter, Jóhannes Kjarval, for a drive there. He borrowed a suitable vehicle, but they had to walk from Dómadalsháls. Guðmundur decided they needed to get stronger vehicles and look for a better route. With the help of mountain-going friends he managed to drive all the way on the 16th of June 1946 and take a picture of the car where the lodge now stands.

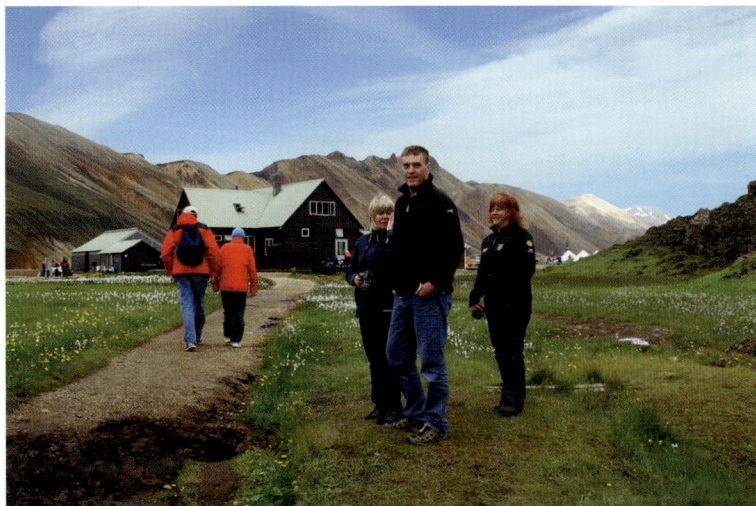

At Landmannalaugar; the Touring Association's main lodge in the background. Photo: Jón G. Snæland

followed by the bathhouse. There is also a little shop, called Fjallafang. The bare necessities have been available there since 1992, with old buses serving as the business premises.

There is little evidence of people living near Landmannalaugar though there are stories of people living near the lake Frostastaðavatn. They also say that Torfi from Klofi brought his folk to live in Jökulgil at the end of the 15th century, to avoid the plague. Thus it was that for fear of outlaws, farmers would not search for their sheep in the area until 1852.

Strútslaug

GPS N 63 52.504-W18 56.677

617 meters asl

The Strútslaug thermal pool is in the Fjallabak region. Most people head for the pool from the Útivist Travel Association lodge, Strúts-skáli, which can be reached by a number of routes. Road no. F120 over Mælifellssandur joins together three approaches from the west. From the east you can take either the Álftakrókur route or one that passes Öldufell. These are passable only by jeeps. The easiest for small jeeps is the Öldufell route, where you enter the highland road near the farm Hrífunes, on road no. 209, avoiding the worst of the fords. Even here you have to ford Brennivínskvísl on the sands by the mountain Mælifell.

All these routes come together by the road between Mælifell and Veðurháls, from which you find the turnoff to Strútsskáli. Over the summers there is a custodian at the lodge who has information about the area.

Resting beside Strútslaug pool. Photo: Jón G. Snæland

The Útivist Travel Association lodge, called Strútsskáli. Photo: Jón G. Snæland

Most people choose to walk the last stretch to the thermal pool, but it is also possible to drive as far as Skófluklif. There is an outhouse there where an old track, called Strútsstígur, begins, and this is a handy place to leave your vehicle. The track leads to the Strútslaug pool and to a ford on the river Syðri-Ófæra at Álftakrókur. A few years ago this track was closed to vehicles, to the dismay of some but the relief of others.

Walk from the outhouse along a gorge and over a ridge called Skófluklif, then to the north of Strútur, where you will find the track. Make your way over the stream and across a saddle between high ridges. Once you are up on the saddle, you will see down into Hólmsárbotnar river bed. From there go in along the slope, over the old driving track that leads down the slope and onto the flat. Innermost to the north-west in the flatland is the Strútslaug thermal pool. The walk takes 1.5-2 hours.

The pool is situated in beautiful, grassy surroundings below the slopes of the Torfajökull glacier. To reach it you have to cross a pretty stream north of the pool. To the south of the pool runs a beauti-

Strútslaug pool in Hólmsárbotnar. Photo: Þóra Sigurbjörnsdóttir

ful river, so the pool itself is on an overgrown spit between the two waterways. The pool has been divided into two, the upper and lower pools, by a rock wall. One pool is 6 by 6 m, the other 8 by 5. The water temperature is 37°- 43°C and the pool is 60cm deep with room for 20-30 bathers at a time. Its surroundings are messily wet, and the bottom is clay, so the water gets cloudy when people bathe there. There are a number of springs around Strútslaug, in one of which a ram once drowned during the autumn roundup, giving it the name Hrútslaug (Ram's Pool). However, there are differing opinions as to exactly which of the springs is the one in question.

Those who wish to continue along the Strútsstígur track will have to cross the clear river, not very deep but fast-flowing near the pool. On the river bank above the pool they have placed a ladder for those that wish to cross there, but it is also possible to cross down on the flat where the river branches out.

Once you have crossed the river you can take a look at an old hut high up on the slope across from the pool. You would hardly notice it if you didn't know it was there, at GPS N63 52.707-W18 55.905. It was built by the German Dick Phillips around 1972 and he has

The memorial plaque at Slysaalda. Photo: Jón G. Snæland

looked after it, living there and using it for his supplies on his many trips to Iceland. The hut has sleeping mats for two. There are many other places of interest near Strútslaug that people tend to pass without noticing.

If you drive across Mælifellssandur to the west you will pass two seldom-used tracks. The first, (GPS N63 48.698-W19 01.519) goes 12 km in along the west of Veðurháls to the gorge Hrútagil. From there it runs along swells and mountains to Svartaklof and then to the mouth of Kaldaklof. Further west is a track starting at GPS N63 49.121-W19 06.165. There you should find some wheel tracks in the sand leading to the swell Slysaalda 700–800 m on (N63 49.437-W19 06.117). This swell got its name (Accident swell) in 1868 when four men died there of exposure. Their remains were not found until ten years later, and bits of their luggage could be seen there, below the swell, far into the 20th century. One hundred years after the accident, in 1968, a descendant of one of the men put up a plaque in memory of the disaster.

Snapadalur

GPS N64 41.451-W17 52.887
960–980 meters asl

The valley Snapadalur is in the middle of the Vonarskarð pass, a large valley between Vatnajökull and Tungnafellsjökull. The valley can be reached either on foot or by jeep. On foot you leave the Touring Association lodge at Nýidalur and follow a marked path up Mjóháls for about 10.5 km.

The hot stream with the mountain Deilir in the distance. Photo: Kjartan Gunnsteinsson

In Snapadalur. Photo: Kjartan Gunnsteinsson

By jeep you can enter Vonarskarð pass from either the south or the north. From the south you turn off road no. F26, Sprengisandsleið, north of Skrokkalda and in as far as the Hágöngulón reservoir. The track becomes less clear as you drive north to the north-west corner of the reservoir, where you have to cross the river Hágöngukvísl. The river can be dangerous, since geothermal heat means it is seldom frozen over. If there is a lot of snow, you may find a snow bridge further upriver. Drive along ridges up the Vonará river, which runs into the reservoir, until you reach a kind of cranny below Kvíslarhnjúkar where the track goes very steeply down to the river. There you need to find a place to ford, but the river is not deep since most water south of the watershed runs along the Kvíslá river bed east of Mt. Skerðingur and Kvíslarhnjúkar. To the north of the Vonará river you reach a wide flood plain where it is safest, in winter, to thread your way north to the ford in the river Kaldakvísl by following the high ground west of the river. To the south of Svarthöfði, however, great care is needed where there is a stream which always has to be crossed on a snow bridge.

Steingrímur J. Sigfússon (now minister of finance) air-dries himself after a long-awaited bathe in the summer of 2003. Photo: Sif Friðleifsdóttir

North of Svarthöfði the Kaldakvísl has to be forded, twice, but these fords are pretty safe. The river is neither deep nor fast-flowing, always open due to a number of hotsprings in Vonarskarð and the fords are usually free of ice floes. From the north of Svarthöfði you can see into the valley Snapadalur.

In winter, head for the valley with Mt. Deilir to your east and Mt. Skrauti to your west. Once past Deilir, drive into a little niche and follow an open warm stream along the hillside. In summer drive over Kaldakvísl and north of Deilir, from where it is only a short walk in to the Snapadalur valley.

Two warm streams converge in Snapadalur. The warmer one is from a spring and has a reddish colour due to the growth of bacteria. The cooler stream also has algae and blue-green algae in it. The cooler stream has been dammed just below a little waterfall, and there is also a dam below where the streams join. The stream is not all that

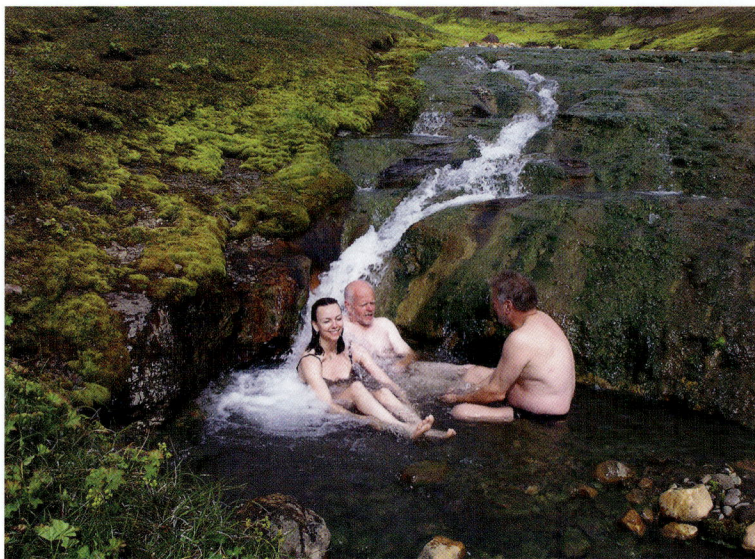

Sif Friðleifsdóttir (MP, former minister for the environment), Kári Krist-jánsson and Þorkell Sigurbjörnsson bathing in Snapadalur. Photo: (probably) Steingrímur J. Sigfússon

good for bathing, with the greatest depth by the falls at only 40 cm. It is about 37°C and runs at a rate of 15 l/sec. The stream has been dammed and re-dammed, and you can't be sure of finding the dams intact in spring.

About 600-700 m to the west of the eastern stream, high on the slopes of Mt. Eggja, is a lovely hot spring field that is not to be missed.

Neither summer nor winter is it safe to drive in the Vonarskarð area in unaltered jeeps or with only one vehicle. In the summer there are dangerous quicksands in rivers and flood plains, and in winter there are dangerous geothermal areas, often under snow.

Few good maps show tracks in the Vonarskarð area, but the best are to be found in the 1:50.000 series from Landmælingar Íslands (The National Land Survey of Iceland). Tracks in the area are also now found in MapSource, the Garmin GPS-driving software.

Hitulaug (Baldurslaug – Volgalaug)

GPS N64 51.677-W17 39.405
783 meters asl

Four km north of the bridge over the river Skjálfandafljót, where road no. F910 branches into Dyngjufjallaleið going north and Gæsavatnaleið going east, is a marked turnoff leading to Hitulaug by the Marteinsflæður flood plain.

Until the bridge was built in 1986 this thermal pool was extremely hard to get to, and is still rather tricky for smaller jeeps, though there are a number of approaches to it. Firstly, road no. F910 leading from Sprengisandsleið (F26) north of Tungnafellsjökull. There are a number of fords on this route which can be dangerous for small jeeps on a sunny summer's day. There is also a route from the valley Bárðardalur and to the east of Skjálfandafljót, but this is very long and slow, with a

Sigurbjörn Magnússon bathing in Hitulaug early one summer morning. Photo: Jón G. Snæland

A group travelling by jeep, at Hitulaug. Photo: Jón G. Snæland

multitude of small, unbridged rivers. From the east one could choose either Gæsavatnaleið or Dyngjufjallaleið (Þríhyrningsleið). Both are rather slow, but there are no unbridged rivers on Dyngjufjallaleið.

The Hitulaug pool, about 3 x 4 m and 50-70 cm deep, with plenty of room for 6 to 8 bathers, is situated in a small loess. About 27°C hot water flows from a crack in the bottom of the pool at the rate of 1 l/sec. The bottom of the pool is gravelly so the water does not get cloudy with use. There are no buildings or changing facilities but quite a bit of vegetation, so it is easy to find a place to leave your clothes.

There are interesting places to visit in the vicinity of Hitulaug, including two good waterfalls. Just west of the bridge over Skjálf-andafljót is a track leading south. About 1 kilometer along it is the waterfall Gjallandi in Skjálfandafljót which is well worth a visit. Not far from Gjallandi is another pretty falls in the clear river Hraunkvísl, whose source is east of Rjúpnabrekkur under the Bárðarbunga icecap. It is also interesting to drive south to the Gæsavötn lakes, about 12 km

The waterfall Gjallandi in the river Skjálfandafljót. Photo: Jón G. Snæland

away, where there has been a lodge since 1973. A newer one was built by the "Friends of the Geese" club, in 1997.

Laufrandarlaug (Hitulaug ytri)

GPS N64 57.970-W17 40.342
739 meters asl

Laufrandarlaug, sometimes known as Outer Hitulaug, is in the western part of the Ódáðahraun lava field, south of the valley Hraunárdalur. Just above Hitulaug (see p. 54) there is a track continuing to the northwest of a swell marked 946. Once north of the swell you have to thread your way through rough lava continuing west of Mt. Steinfell. Just under three km beyond Steinfell take a left at GPS N64 56.339-W17 39.165 to enter a track leading down to Hraunárdalur, reaching Laufrandarlaug in about 7 km. Those who plan to go to the North can continue along Hraunárdalur for another slow 32 km to road no. F26. Or you can visit the pool and return to the shorter route to the North, called Urðir and Laufrandarleið, leading to the eastern side of Bárðardalur valley.

The Laufrandarhraun lava field is also known for being the main nesting ground in Iceland of the snowy owl, an impressive bird which can reach a length of 66cm and weigh up to 1.9 kg. It nests mostly in tundras and heaths and eats goslings, ptarmigan and waders.

It is inadvisable to travel on your own in the wilderness area east of the river Skjálfandafljót. There is not much chance of running into other travellers north of Hitulaug (p. 47), so it is important to be well equipped, ideally with a GPS direction finder, good communications, and driving an altered jeep with tyres of at least 33-35", since the tracks are often rockbound and crossed by narrow gullies.

The only shelters are the Gæsavatnaskáli hut by Dyngjujökull in the south and the Réttartorfuskáli hut near Hafursstaðir in the north. Both are kept locked and need to be booked ahead; Gæsavatnaskáli with Gæsavinafélagið ("Friends of the Geese" association) in Akureyri, Réttartorfuskáli with the Eyjafjörður branch of the 4x4 Travel Club.

Guðmundur Gunnarsson describes Laufrandarlaug in The Touring Association Yearbook of 1981 as follows:

The thermal pool itself is made up of a number of springs in a dell by the lava edge. They vary in temperature, the hottest being 40-50°C.

There is a small, marshy vegetated area around them where the water runs in rills down to a sizeable stream which is formed partly by the springs from under the lava and partly from the so-called Hitulau-gardrag sloping up to the gravel moor south of the pool. After the stream has collected all this water, hot and cold, it leaves the lava's edge and merges with another stream from Langadrag that flows a few kilometers further west, flowing finally into the Hrauná river. The pool is not the only geothermal water in the area. There are signs of hot water here and there, both in the vicinity of the pool and further off on the gravel moors to the west and south. They are often about 20-30°C.

Grímsfjall

GPS N64 24.413-W17 15.960

1732 meters asl

The highest thermal pool in Iceland is on the mountain Grímsfjall, near the middle of the Vatnajökull glacier, south of the Grímsvötn caldera. The most active volcano in Iceland is in this mountain which is 1732 m asl. Members of the Iceland Glaciologial Society built a lodge there in 1957, and again in 1987, when a building containing showers, steam-bathing and toilet facilites, a storeroom and fuel storage was also added.

The surroundings are among the most majestic in Iceland. The buildings are on the summit of Svíahnúkur Eystri and the view from

The Iceland Glaciologial Society lodges on Grímsfjall. Photo: Kjartan Gunnsteinsson

Grímsvötn

A caldera near Grímsfjall

There are not many older written records of Grímsvötn. There is, however, an old folktale told to Árni Magnusson by a man who had been born in 1680. It tells of Grímur of the Western Fjords, who had killed a man. He looked to a woman in the north for sanctuary and she told him of good fishing lakes to the south where he could hide out until he could make his escape by sea. Grímur found the lakes, surrounded by woods and with plenty of fish. He came across a giant by the lakes, killed it and then married the giant's daughter. She prophesied that the lakes where Grímur dwelt would from time to time flame up and burn away the woods that surrounded them. This has often come true since.

the roof on a sunny day is incredible, and makes you feel you must be at the very top of the country. Among many others you can see Lómagnúpur, Mt. Þumall, Öræfajökull, the Esjufjöll mountains and many more. Just north of the lodge is the 300 meter sheer drop called Stórkonuþil, onto the ice covering Grímsvötn, with the Griðarhorn peak a bit further east. To the west is an extensive geothermal area and even further west is Svíahnúkur. To the south and west of the highest point is a caldera and geothermal area that has been growing in the last few years, making it very dangerous. People used to drive through there onto the glacier, but now they have to detour round it.

There are several routes for driving up to Grímsfjall, the most common being from the Jökulheimar lodge. The drive is about 37 km:

The snow-covered steam-bath at Grímsfjall. Photo: Kjartan Gunnsteinsson

across the river Tungnaá, which is usually just a frozen trickle in the winter, and up the fairly gently sloping Tungnaárjökul glacier. People often take a detour via Mt. Pálsfjall, about 6-7 km south of the direct route to Grímsfjall.

Hveragil

GPS N64 41.677-W16 30.387

901 meters asl

Hveragil is north-east of Kverkfjöll and east of Rauðutindar, in an angle between Kverkjökull to the east and Skarphéðinsjökull and Brúarjökull to the west. It can be reached along a track passable by all jeeps.

Approaching from Hvannalindir you drive along a good track til you reach the mountains of Kverkfjallarani where you take the 26 km summer track (GPS N64 50.266-W16 24.396) to Hveragil. The least-defined parts of the route have been marked with stakes.

In a dramatic gorge, hot water joins with a small clear-water stream to produce a comfortably warm bathing temperature. There are two really good bathing places: one is a pool under a waterfall, the other a little way above the falls. This stream is thus unique among thermal pools in Iceland.

In Hveragil. Photo: Kjartan Gunnsteinsson

At the mouth of the famous Kverkjökull ice cave. Photo: Jón G. Snæland

Kverkhnjúkaskarð pass where you join the track west of Kverkfjöll to the lodge. Sigurðarskáli is a large building which sleeps 85. It also has recently-built bathing and toilet facilities, a camping ground in the summer, and a warden who has information about the area. The track continues on for another 5 km to the mouth of an ice cave in the glacier Kverkjökull.

From a distance the Kverkfjöll mountains are a majestic sight. The highest peak to the east is 1936 m asl, to the west 1799 m asl. Between these peaks the Kverkjökull gacier tumbles down in the angle between two high cliffs. Kverkfjöll is a central volcano which has erupted at least 20 times since settlement. There are numerous geothermal areas there, the best-known being Hveradalur west of the mountains. Many travellers who come to the area walk onto Kverkjökull along the so-called Langafönn (long snow) or, in the winter, drive there. However, it is important to note the Kverkjökull has many fissures and can be very dangerous, though people do occasionally drive onto it when the snow situation is judged to be favourable.

The Iceland Glaciologial Society has a hut on Tunnusker in the western part of Kverkfjöll at 1700m asl. Just east of the hut is a great caldera called Gengissigið. It is about 600m across and 100m deep and is thought to have been formed in 1959, when there was some kind of explosive geothermal eruption. The main route south onto Vatnajökull lies to the west of this caldera.

Víti

GPS N65 02.825-W16 43.408
1053 meters asl

The explosion crater Víti is in the Dyngjufjöll mountains just north-east of Öskjuvatn lake. It is perhaps the most alluring place in the Askja area. The crater, which was probably created in a steam explosion towards the end of 1875, is about 60m deep and 300m across. The water in it reaches a depth of 8 m. It is sulphuric and varies greatly in temperature, from 22°-60°C. It can be dangerously hot in places, especially in the mud of the eastern shore. There are no changing facilites near Víti so the fun is in bathing as Adam and Eve did, a sport that foreign visitors seem to appreciate.

The most direct approach to Víti is from the north, along Highway 1, taking the turnoff onto road F88, Öskjuslóð, just west of the bridge over Jökulsá á Fjöllum. This gravel road to Askja is about 100 km long and passable by most jeeps, so long as the water is not too high

The explosion crater Víti, Öskjuvatn lake in the background. Photo: Guð-mundur Einarsson

Caution is needed since there may be hidden hot springs on the bottom of Víti. Photo: Sigurbjörn Magnússon

in Lindá by Herðubreiðalindir. It can also be reached from the east or the west along road F910.

When you reach the Dyngjufjöll mountains on road F88, the first thing you see is the Drekagil gorge, which is very striking. There are some good service lodges in Drekagil, along with some Touring Association lodges. From there you take the F894 uphill to Öskjuvatn lake, ending at a parking area by the mountain Biskup. From there it is about a 35 minute level walk to Víti. Once there you have to descend a steep slope which can be very slippery when wet, so caution is needed.

Nearby is Öskjuvatn, a 217 m deep lake that was created in an eruption in 1874-75. If you plan to spend a few days in the Dyngjufjöll area there are various interesting places to explore.

Heading south from Drekagil along the F910 towards the mountain Vaðalda you come to a road sign marked Svartá. From there it is a short drive to the source of the Svartá river which comes spurting suddenly out of the coal-black sands and is immediately deep. Further

The path down to the water in Víti is steep. Photo Óskar Erlingsson

west, just before Svartá joins Jökulsá á Fjöllum, there is a lovely water-fall, called Skínandi ("Shining").

In Drekagil. Photo Jón G. Snæland

Continuing along the F910 beside Jökulsá á Fjöllum you reach an area called Flæður, or Síðdegisflæður ("The afternoon flood plain"). The river beds are dry there early in the day, but the afternoon brings a flood-wave from Dyngjujökull. Around there you can either turn south over the flood plains or continue west and north towards the dramatic landscape of the valley Dyngjufjalladalur, keeping right at the next fork in the road. The Touring Association has a small hut there which is just right for a coffee break. As the valley opens out you get a splendid view of the lava fields Útbruni and Frambruni.

In the right vehicle with the right equipment and guidance, it is possible to follow a complicated route from here north and east of the Dyngjufjöll, making a circle route back to Drekagil.

Guðlaug

GPS N64 09.651-W17 29.407

821 meters asl

The thermal pool Guðlaug is on the banks of the river Bergvatnsá in the valley Beinadalur south of Vatnajökull. The valley is 2-3 km west of the glacial lagoon, Grænalón, which has icebergs floating in it – or sitting on the dry lake bed after a glacial flood. This lake is one of the deepest in Iceland, up to 200 m at its fullest, when its area can be 18 km². When it floods, the water flows to the river Súla, sometimes increasing its flow to 2000 m³ per sec. The river Bergvatnsá, on the other hand, is one of the sources of the Núpsá river, which flows to the lowlands and into the lovely Núpsstaðarskógur woods.

The Guðlaug pool has room for 5-7 bathers but is only 20-30 cm deep. It could be deepened by slightly raising the wall where the run-off flows into the river. It is usually approached along the Beinadalur

Walkers rest their weary bones in Guðlaug in the Beinadalur valley. Photo: unknown

valley alongside Núpsá or Djúpá, a full day's walk, with some rivers to wade across. It is possible to drive from Vatnajökull down to Grænalón, usually off of Skeiðarárjökull but possibly off the glacier west of Geirvörtur in the right conditions. In winter it may be possible to drive from Miklafell, crossing the river Hverfisfljót, then along the south of Síðujökull and up Langasker.

The West

Breiðafjörður

Stykkis

Grundarfjörður

Stjáni
Sigga ★ Lýsuhólslaug

Faxaflói

The West

Pools are marked with a star. Pools
marked with a green star are not
specifically covered.

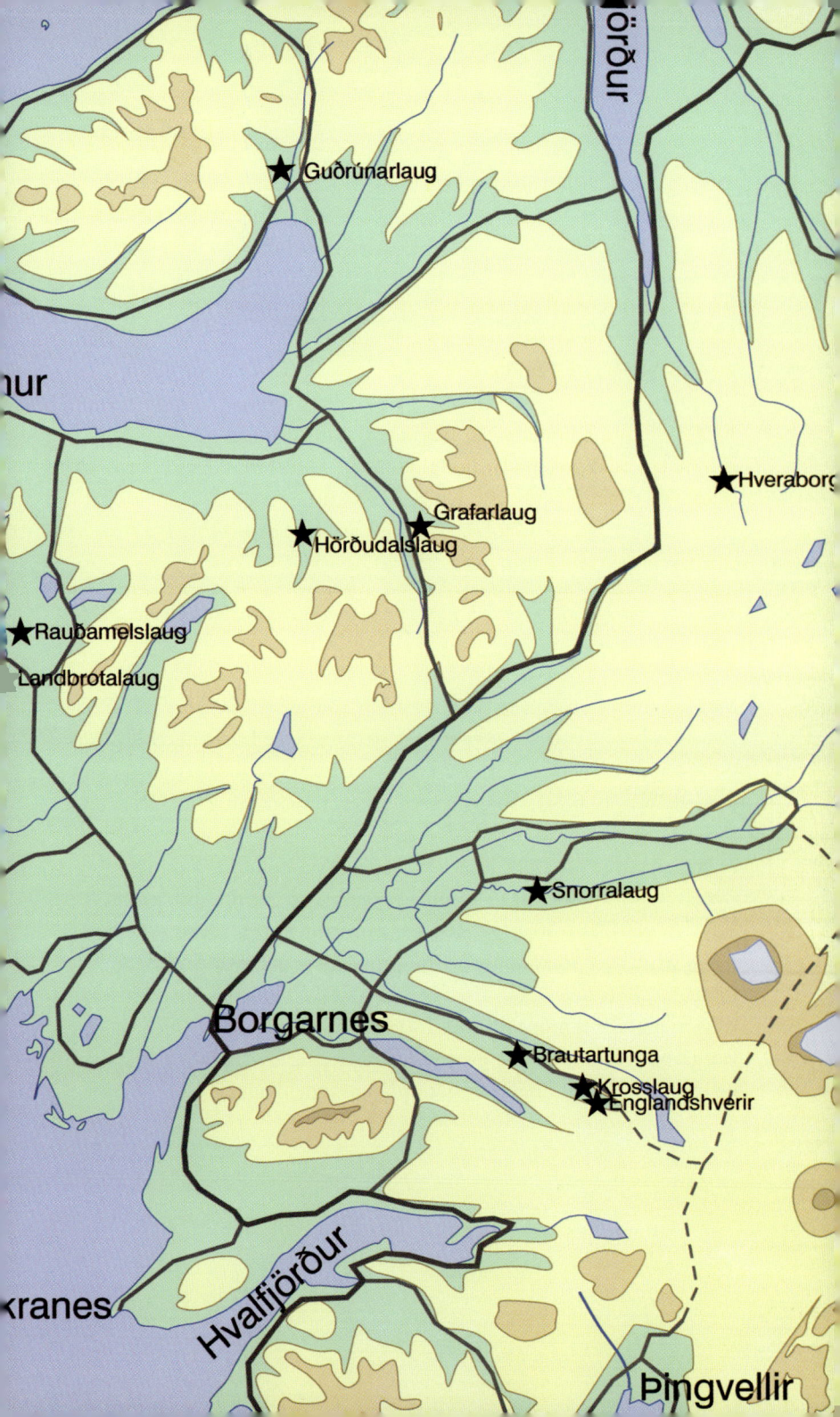

One can cross the river on this pipe which houses the hot water hose. Photo: Jón G. Snæland

by the pump house. The pipe is strong and is easy to cross. Near the river bank you can see where two streams, one hot, one cold, were once dammed with a fair-sized turf and stone wall. This pool (N64 29.539-W21 10.600) was once used to teach swimming but has grown too cold.

If you walk up along the hot stream you come to a spring from which 90°C water runs through a pipe over to the pump house (GPS N64 29.465-W21 10.645). There the run-off forms some interesting pools which are too hot for bathing but further down the stream the temperature is just right. The farm England is no longer lived in year-round, but it is famous for the story of a farmer who was once visiting Borgarnes. When people expressed surprise at the name of his farm, he insisted, "Well, England isn't only in Copenhagen!"

Krosslaug (Reykjalaug)

GPS N64 30.244-W21 12.235

130 meters asl

Krosslaug, sometimes called Reykjalaug, is on the farm Reykir in Lundarreykjadalur valley. It is about 50 m above the road and the temperature is 42°C.

In *Kristni saga* (The story of Christianity) it says that in the year 1000, when Christianity was declared the national religion, most of the representatives at the Althingi did not want to be baptised in cold water. Those from the north and south were baptised in Reykjalaug (Vígðulaug) at Laugarvatn, while those from the west were baptised in Reykjalaug in "Southern Reykjadalur", which may have been Krosslaug in Lundarreykjadalur or a thermal pool nearby.

Krosslaug is in a lovely clearing just above the highway. Photo: Jón G. Snæland

Reykjahver hot spring, with a temperature of 75°C, is also on the farm Reykir. In the middle of the 19th century they built a swimming pool there. Both the pools are well-preserved.

Brautartunga

GPS N64 31.805-W21 19.696
39 meters asl

Brautartunga is in the Lundarreykjadalur valley, easily reached by all vehicles along roads 52 or 512. The Youth Association Dagrenningur, established in 1911, runs a big social centre there that was built in 1946. The centre, which can seat 250 and also has sleeping bag accommodation, is available for rental to groups. Beside the centre is a 12.5 x 6 m concrete swimming pool, built around 1940, with two changing rooms. The water comes from the Brautartunga hot spring north of the river Tunguá.

The hot spring comes up from under a gravel ridge on the river bank a short distance below the farm. The water is over 90°C. When the Tunguá river was in spate it used to overflow into the hot spring and cool it, but now the river bed has been moved away from it. The Reykjahver hot spring, about 76°C, is also at Brautartunga, right beside the farm house.

The Brautartungulaug thermal pool, built around 1940, is of a size that was common at the time. Photo: Jón G. Snæland

Snorralaug

GPS N64 39.842-W21 17.473 WGS 84

38 meters asl

Snorralaug is in Reykholt in Borgarfjörður county, a few kilometers inland from the Kleppjárnsreykir hot spring. It is named for Snorri Sturluson, 1179–1241, though it is mentioned in records from long before his time. In *Landnáma* (*The Book of Settlement*, written about 1130) it says there was a thermal pool at Reykholt from the year 960. The pool, which is 4 m across and 0.7-1 m deep, has water piped into it from a nearby hot spring called Skrifla. The pool has been rebuilt several times through the ages, for instance in 1858 by Pastor Vernharð Þorkelsson, whose initials (V. th.) can still be seen on a stone in the pool wall. It has since been rebuilt twice, in 1959 and 2004, by the Archeological Heritage Agency of Iceland.

Snorralaug and the small house that was built to protect the tunnel which once led to the farmhouse. Photo: Þóra Sigurbjörnsdóttir

The tunnel that was said to lead to the farmhouse and could be used as a means of escape if enemies suddenly attacked the bathers. Photo: Þóra Sigurbjörnsdóttir

Snorralaug was one of the first (1817) officially protected archeological sites in Iceland. Beside the pool is a small house, built over a tunnel that is said to have connected the farmhouse to the pool in

The Killing of Snorri Sturluson

In *Sturlunga Saga* they describe how Snorri was killed in 1241.

Gissur and his men searched the buildings, looking for Snorri. Gissur found Father Arnbjörn and asked him where Snorri was. He said he didn't know. Gissur said he and Snorri could not make peace if they did not meet. The priest said he might be found if he were promised a pardon.

After this they became aware of Snorri's whereabouts. And they went to the cellar, Markús Marðarson, Símon knútur, Árni beiskur, Þorsteinn Guðinason, Þórarinn Ásgrímsson. Símon knútur told Árni to strike him. "You shall not strike," said Snorri. "Strike him," said Símon. "You shall not strike," said Snorri. Then Árni gave him a death blow and both he and Þorsteinn attacked him.

the time of Snorri. The building is not a replica, but simply meant to protect the tunnel from the elements. It is also said that the tunnel was intended as a means of escape if Snorri's enemies attacked, and that he indeed tried to flee through the tunnel to the cellar when he was killed on September 23rd 1241.

Landbrotalaug

GPS N64 49.933-W22 19.110

23 meters asl

This pool is on the farm Landbrot on the southern side of Snæfellsnes, just south of the river Haffjarðará. Approaching along road no. 54 from the east, about 1.6 km before crossing Haffjarðará, turn off to the left along the road to Stóra Hraun. Cross a cattle grid onto a good gravel road. In about 1.5 km you pass the abandoned farm Landbrot, and about 150 m further on turn left again onto a faint track. About 300 m further on there is a parking area from which a faint path leads to the pool. You have to cross a small and very shallow pond over stepping stones that have been placed in it, or simply wade across.

The Landbrotalaug pool is south of the pond, with a stone wall around it. Its surface is about 1 x 1 m, its depth 1.5-1.6 m and its temperature 33-35°C.

Mother and daughter taking a footbath in Landbrotalaug. The pond that must be crossed is in the background. It is shallow enough to cross dry-footed on stepping stones. Photo: Jón G. Snæland

Wading in the pool by the borehole just south of Landbrotalaug. The mountains Kolbeinsstaðafjall and Fagraskógarfjall in the distance. Photo: Jón G. Snæland

There is another thermal pool south of Landbrotalaug. To get there you take a left from the parking space along another faint path to the pond. Wade across and you come to a borehole, where a dam has created a small lagoon. The water is shallow, but the bottom is gravel, so you can lie flat to get immersed.

Rauðamelslaug

GPS N64 52.206-W22 17.021 WGS 84
55 meters asl

Rauðamelslaug is an ancient turf pool. It is approached from road no. 54 from which you turn off onto no. 55, which goes through the Heydalur valley. Past the Rauðhálsahraun lava field take a left and drive north of a crater along an all-vehicle road for about 6.3 km, til you reach the farm Syðri-Rauðamelur. There you turn and head for Syðri-Rauðamelskúla, where you will find the pool below the south-east side. A couple of dozen meters above the pool is a borehole with a shed that can function as a changing room.

There are two thermal pools there, very close together. The larger pool is 3 x 7 m with a temperature of about 40°C. It has a mineral spring in it and is very cloudy with a lot of algae. The lower one is called Guðmundarlaug, named for one of two Guðmudurs: a farmer who lived there and had great faith in the healthfulness of the pool, or Guðmundur the Good, bishop at Hólar in the early 13th century.

Guðmundarlaug is an ancient pool that is mentioned in Sturlunga Saga.
Photo: Þóra Sigurbjörnsdóttir

Above the pool, the borehole shed is convenient for use as a dressing room.
Photo: Þóra Sigurbjörnsdóttir

These pools are mentioned in Sturlunga Saga, where it says that Aron Hjörleifsson came there in 1222. Sturla Sighvatsson had had Aron banished and he hid for a while near Rauðamelslaug, in a cave (now lost) called Aronshellir.

Lýsuhólslaug

GPS N64 50.478-W23 12.855 WGS 84
27 meters asl

The school and pool, Lýsuhólsskóli and Lýsuhólslaug, are about half-way along the southern Snæfellsnes road. The school also functions as a community centre, and the swimming pool is at one end of it. It is known particularly for the fact that it is fed from a borehole containing mineral water and not mixed with chlorine, so it is strictly speaking a natural thermal pool. In other ways it is comparable to standard public swimming pools, and has changing rooms and showers for both sexes as well as a hot tub, 38°C, and a steam bath.

In the summer (June 1 – August 15) the pool is open daily from 13.00-19.00. It is about 16 x 8m, 0.9-2.1 m deep.The pool is emptied and cleaned every week and rinsed with a high-pressure hose. The forerunner of this pool was a rock-walled pool on the banks of the river Lýsa, which flooded so severely in 1980 that it broke down the old pool.

Lýsuhólslaug is a modern swimming pool, but is fed with mineral water and not mixed with chlorine, so it counts as a natural thermal pool. Photo: Jón G. Snæland

Sigga

GPS N64 50.497-W23 13.027 WGS 84

34 meters asl

The hillock Lýsuhóll is a protected site. The name means the bright hillock. There are two thermal pools there, Sigga and Stjáni, close to each other. Sigga, which has been hewn in the silica, is small and shallow, with a great deal og algae. It is hardly used at all since the advent of Stjáni. Sigga is thought to be very old, and certainly *Þorgils saga Skarða* mentions a pool on Lýsuhóll, though it was probably not Sigga itself, as it was larger. The Saga speaks of a pool at Lýsuhvoll:

> One day Þorgils rode to Lýsuhvolslaug for his entertainment and his companions with him. Vestarr Torfason dwelt there. Jóreiður was his wife and she was handsome. Þorgils and his companions dismounted by the pool. Jóreiður was there before them, washing clothes. Þorgils took her hand and joked with her. Her husband disapproved. He took up his weapons and there was almost a fight to the death.

Sigga often has a lot of vegetation and isn't tempting if it has not been cleaned for a while. Photo: Jón G. Snæland

Stjáni

GPS N64 50.497-W23 13.027 WGS 84

33 meters asl

The farm Lýsuhóll is left of the road from 54 to Lýsuhólsskóli. Turn left at the road to the farm til you come to a strong gate. From there walk a few dozen meters to a light-coloured hillock, where you can find the thermal pools Sigga and Stjáni. The latter is a plastic tub that has been buried so only 20 cm stand up from the ground. The tub is 2 x 2 m and water comes through a plastic pipe that rests on the edge. The water runs at about 1 liter/sec and the temperature is about 45°C. Both water and tub are clean, but the surroundings are rather wet and muddy, so clothes must be left at quite a distance on the grass.

The pool Stjáni is actually a tub, but this is hard to see because of the calcium deposits on its edges. Photo: Jón G. Snæland

Hörðudalslaug

GPS N64 57.225-W21 44.607 (approximate)
70 meters asl

The pool is in the valley Hörðudalur (road no. 581) south from road no. 54 near the inner end of the Hvammsfjörður fjord. Once past the farm Hlíð and the bridge over Hörðudalsá fork right past a sheep corral, then turn left onto a jeep track into the valley Laugardalur. The ruins of a cement building can be seen below the road a few hundred meters further on. They are what is left of the dressing rooms by a swimming pool that was built by the local swimming club in 1930 but is now unusable due to frost damage.

There had been a stone-built pool there since 1915, and when they started digging to lay pipes to the new pool they found a stone-built circular pool, probably ancient. There is another pool above the road which you can find by following a warm creek upstream past the ruins. It is stone-built up against the grass slope. It has not been kept

The old Hörðudalslaug pool could use the help of some handy and hardworking people to again become a pleasant little hot pot. Photo: Jón G. Snæland

The Hörðudalslaug pool was impressive years ago, and had a large building attached, which is collapsing since the pool was damaged by frost. Photo: Jón G. Snæland

up, and at some point they laid a metal pipe right across the pool. But the temperature is good, and with some work, the pool could be made an attractive one for 2-3 bathers.

A jeep track continues in the valley, which is quite beautiful and worth exploring. It then goes up rather steeply over Sópandaskarð pass to the Langavatnsdalur valley, leading to lake Langavatn, then over Beilárháls ridge and along road no. 553 to Highway 1 at Svignaskarð; 40 km. This route has become increasingly popular, both for jeeps and horses.

Grafarlaug (Reykjadalslaug)

GPS N64 57.647-W21 30.953
100 meters asl

Just to the north of the farm Gröf on road no. 60 turn off and drive past the Grafarrétt sheep corral in the Reykjadalur valley. The track is passable by all vehicles, but rather rough since the fill is taken from the gravel banks of the Reykjadalsá river. It is about 2.3 km to the pool, which was built by the local youth association in 1956. The pool may not have water in it – it has not been kept up during recent years, though there are plans to do so.

The pool is concrete, 12.5 x 4.9 m, with a depth of 0.9-1.3 m. The water is usually 26°C but is said to sometimes go up to 80°C, so people are warned against leaving children alone in the pool. The changing rooms were built more recently. They are divided in two, so the sexes can change separately, but they have no doors, and the grassy surroundings are fine for keeping clothes on. The water is from a borehole in the building of which an old stone-built pool from the time of Sturlunga Saga was unfortunately destroyed.

Grafarlaug in the Reykjadalur valley. The distant mountain is Sauðafell in Miðdalir, and below it is the farm Fellsendi. Photo: Jón G. Snæland

Guðrúnarlaug

N65 14.777 W21 48.304
Í 90 metra hæð yfir sjávarmáli

This pool, in the valley Sælingsdalur in Dalasýsla county is named for Guðrún Ósvífursdóttir, one of the great heroines of the Icelandic Sagas. Rededicated in October 2009, 140 years after it had been lost under a landslide, the new Guðrúnarlaug was built near the site of the original pool, and is fed by the same spring. Guðjón Kristinsson, a stone-mason from Drangar in the Strandir area, built the pool. It has stone-built underwater benches along its walls. A short way above the pool they built a so-called "modesty house", or changing room for the shy bather.

The modesty house is built in the ancient style, with stone-built walls and wooden gables. This pool has a strong presence in the ancient sagas and is mentioned in both Sturlunga Saga and Laxdæla Saga. One of the main characters in Laxdæla, Guðrún Ósvífursdóttir, was the daughter of Ósvífur Helgason, son of Óttar, son of Bjarni the oriental, son of Ketill flat-nose. Her mother was Þórdís daughter of Þjóðólfur the short. Guðrún grew up on the family farm at Laugar in the Sælingadalur valley.

Guðrúnarlaug in Sælingsdalur. Photo: Steingrímur Steinþórsson

The Westfjords

Laugaland

GPS N65 30.709-W22 18.189 WGS 84

13 meters asl

Laugaland is near the mouth of Þorskafjörður on the southern side. You leave road no. 60 and take 607 past Reykhólar and on around the peninsula til you reach the farms Staður and Árbær. The road turns from there down to a small harbour. In the bend you will see a gate on your right, and a track beyond it. Go through the gate and drive 5 km to the farmhouse at Laugaland. This is a jeep track, but should be passable by all sizes of jeeps. The drive is beautiful, below steep mountainsides along a shore called Hákarlaströnd (Shark Shore). When you reach the farmhouse you will see the pool on your left, in a dell just inside the fence.

The pool is concrete, 7.8 x 3.8 m and 0.6-1 m deep. About 0.2 l/sec run into the pool through a pipe. The water is about 49°C. There are

The pool by the farm Laugaland. Photo: Jón G. Snæland.

Jökulfirðir

Bolungarvík

Ísafjörður

Ísafjörður

Dýrafjörður

Rey
Gamlalaugin Reyk

Galtarhryggslaug

Arnarfjörður

Dynjandislaug

Brúarpottur
Pollurinn

Reykjarfjarðarlaug

Vegavinnubaðið

Patreksfjörður

Hellulaug

Mórudalslaug

Krosslaug

The Westfjords

Pools are marked with a star.
Pools marked with a green star
are not specifically covered.

Breiðafjörður

Drangajökull

★ Hestvallalaug

★ Krossneslaug

★ Hákarlavogur

Laugarás

Nauteyrarpottur
★ Nauteyrarlaug
★ Rauðamýrislaug
arfjallslaug
aug
vottalaug
alur

Húnaflói

★ Gvendarlaug

Hveravík
★ Drangsnespottar
Steingrímsfjörður

★ Laugaland

Hrútafjörður

The farm Laugaland by Þorskafjörður. On the other side of the fjord is the peninsula Hallsteinsnes where the Teigsskógur woods are. There has been controversy over whether to build a road through the woods. Photo: Jón G. Snæland

no changing facilities and the surroundings are quite overgrown, not all that convenient for laying clothing. There is a pool on the shore, called Prettur. It becomes accessible at low tide.

Vegavinnubaðið

GPS N65 37.601-W22 56.991 WGS 84
223 meters asl

Vegavinnubaðið (The Roadworks Bath) is at the end of the fjord Kjálkafjörður, so high up on the slope that you need to be fit to cope with the walk. You can leave your vehicle at the bridge over Kjálka-fjarðará and follow the river up stream on the east side, along an old track. When the track gives out, continue alongside the river to a gorge, which cuts across your path. Cross the gorge and climb up the northern side of it, then continue up the slope with the river on your left. There are a number of waterfalls along the way, two of which are very impressive. When you reach a height of 223 m you will be very close to a big waterfall in the river. In the cliff a bit below the falls on the eastern side hot water flows out of the rock, about 8-10 l/sec at a temperature of 25-27°C.

At one time the water was collected in a wooden channel to form a shower at a height of about 1.5 m. All that is now left of this construction is two bits of wood trapped under boulders that have fallen down

Here we see a great deal of warm water flowing out of the cliffside. Photo: Jón G. Snæland

The Kjálkafjarðará runs down the colourful slopes in innumerable waterfalls.
Photo: Jón G. Snæland

the cliff. It may well have been a roadworks crew, who made the road over the Þingmannaheiði heath at the fjord's end in 1950-5, that built the shower.

Hellulaug

GPS N65 34.637-W23 09.579

1–2 meters asl

The pool is 4-500 m in from the Flókalundur Service Centre. From road no. 62 you can drive down towards the pool to a well-marked parking area. At the end of the parking space is a borehole where there once was a pool, and from there a plastic hose carries water down along a bank that hides Hellulaug itself.

The pool is on the seashore below the bank and is more easily reached from the eastern side. There you can lay your clothes on the rock, as there are no changing facilities. The pool is situated in a hollow or inlet in the rock. A pool of 3 x 4 m has been formed there, built of both rock and cement. It is about 60 cm deep and the temperature around 38°C. The gravel on the bottom is very fine, so it clouds up a

Hellulaug by Flókalundur. Photo: Jón G. Snæland

On Þingmannaheiði heath; Vatnsfjörður in the background. Photo: Jón G. Snæland

bit. Hellulaug has, like the islands and beaches of Breiðafjörður, been an officially protected site since 1995. It is tended by the employees at Flókalundur.

The services at Flókalundur include one of the better camping grounds in Iceland, and there is a variety of interesting activities in the area. Hellulaug is a perfect walk away, and Þingmannaheiði heath is nearby. The oldest route across it is marked by well-built cairns, but the 1950 bull-dozed road, which starts just near Flókalundur, has not been kept up since 1969, and was closed pretty much entirely in 1974. It is, however, possible to drive a short way along it in altered jeeps with 33" or larger tyres, and the countryside is lovely with waterfalls, gorges and woods. Those who don't feel up to walking can take four-wheeler trips which are on offer at Hotel Flókalundur.

Flókalundur is named for the Norwegian Viking Hrafna-Flóki Vilgerðarson. In the Book of Settlement it says that Flóki dwelt for some months in Vatnsfjörður in the year 865. "Then Flóki climbed a

One of the carefully built cairns that mark the old route across Þingmanna-heiði. Photo: Jón G. Snæland

high mountain and looking north of the mountain saw a fjord full of icebergs, so they named the land Iceland." This was the first recorded mountain climbing in Iceland, and Flóki and his men are thought to have climbed the mountain Lónfell, north-east of Vatnsfjörður.

Krosslaug

GPS N65 31.178-W23 24.335 WGS 84
On the seashore

Krosslaug is on the seashore, on a promontory called Lauganes below the Birkimelur settlement by the Hagavaðall inlet. The swimming pool can be seen from the path leading from road no. 62 where it passes Birkimelur down to the shore. The pool was built by the local youth association in 1948 and swimming lessons started there in 1949. The concrete pool is 13 x 5 m, 1-2 m deep, the water temperature around 35°C. A few years ago, when the pool and its facilites had rather deteriorated, the community built new changing facilities and renovated the pool and its environs.

Krosslaug before its total renovation. Now the fence has been pulled down and new dressing rooms built. Photo: Jón G. Snæland

Mórudalslaug

GPS N65 32.803-W23 25.180 WGS 84

77 **meteters asl**

The Mórudalur valley is inland from the settlement at Birkimelur and the Hagavaðall inlet. The road to the Mórudalslaug pool lies from road no. 62 and along the valley, just beyond the turnoff to the farm Kross.

The track in the valley is 3 km long, passable by most small jeeps, but not by saloon cars. You have to cross a rocky stream, along a rather rough track, and finally ford the Mórudalsá river which is, however, not deep in dry weather and has a fine gravelled riverbed. The track ends on the gravel bank of the river and there you leave your vehicle to walk up along the east side of the river and into the woods along its banks. Soon you come to a little stream (GPS N65 32.770-W23

The Mórudalslaug thermal pool, with little water and lots of plant life. Photo: Jón G. Snæland

25.388) where there is a little cairn to show the way. Walk up along this stream for about 700 rather difficult meters. This is about a 15 minute walk.

The pool itself is on a lovely green slope. It has almost no water in it, but a hot stream runs through it, 30-35°C. The pool was built of turf and rock in 1903, probably by the people at Kross. It is about 10 x 5 m, surrounded by a wall that is about 1 m high and 1 m thick. They taught swimming there from 1903, probably for a full 45 years, or until Krosslaug appeared. It would be fairly easy to dam this pool again; more work, but worth it, to clean out the vegetation and arrange for run-off.

Brúarpotturinn

GPS N65 39.411 W23 54.618 WGS 84

8 meters asl

Brúarpotturinn (The Bridge Pot) is in Tálknafjörður fjord, 4–5 km beyond the village Tálknafjörður. The pool is on the western bank of the river Laugardalsá, almost under the bridge itself. A concrete wall has been built by a little rift in the rock to make a pool with room for 4-5 bathers. Water at a temperature of about 36°C is led through a plastic pipe from a borehole further up the bank. The pool, which is privately owned, is on the farm Litli-Laugardalur.

The next farm along is Stóri-Laugardalur, where there is a 100-year-old church, formerly the Tálknafjörður parish church. According to

Children at play in Brúarpotturinn. Photo: Jón G. Snæland

The river Laugardalsá empties into Tálknafjörður. Photo: Jón G. Snæland

the town's website the building material was imported from Norway. The impressive pulpit, which is very old, is thought to have been formerly in the Cathedral in Odense, Denmark, birthplace of H.C. Andersen. The gilded chalice is very old, and one of the church bells is inscribed "Torolfer Ulafsen, anno 1701".

Pollurinn

GPS N65 38.945-W23 53.669 WGS 84

34 meters asl

Pollurinn, or "The Puddle", is not far from Sveinseyri, west of Tálkna-fjörður village on the way to Stóri-Laugardalur. It is just above the road, but easily missed as you drive past. It consists of three concrete pools: one 2 x 2 m with a depth of 1 m, another 1.35 x 2 m with a depth of 0.5 m and the third is 2 x 2.8 m with a depth of 0.4 m. The water temperature is about 46°C. There are good facilities, for both men and women, right by the pools for changing, hanging your clothes and showering.

They drilled for hot water there in 1997 and found enough hot water for the swimming pool at Sveinseyri, and to heat the school and gymnasium. The local authority sees to maintenance.

Pollurinn is extremely popular with the citizens of Tálknafjörður. Photo: Guðlaugur Jakob Albertsson

Reykjafjarðarlaug

GPS N65 37.381-W23 28.129 WGS 84

3 meters asl

Reykjafjarðarlaug is right beside the highway at the end of Reykjafjörður, off Arnarfjörður. There are actually three pools here, one of
which is beside a fenced-off summer cottage and is privately owned.
The concrete main pool, a full 16 x 10 m with 32°C water to a depth
of 1.2 to 1.8 m, was built by volunteers in 1975. The local authority
provided building materials, the landowners provided the land and
hot water.

Originally it was agreed that the local authority would be owners
for the first 10 years, and after that the land-owners. As things stand
today, the local authority claims there is a camping ground by the
pool, but there are no facilities, and exactly who now owns the pool is
anybody's guess.

However, the pool, which is open all year round, is usually clean
and well-kept-up. There is an old road workmen's shed with two

*The old pool in Reykjafjörður, above the concrete pool. Photo: Jón G. Snæ-
land*

The main pool in Reykjafjörður, built in 1975. Photo: Jón G. Snæland

changing rooms, which is nearing collapse, but provides some shelter. There are some flagstones alongside the shed and part of the south end of the pool.

A few dozen meters above the main pool are two natural pools, but the lower one, which was turf-built, no longer has water in it. The upper one is a warm stream with a stone-built dam. It is diamond shaped, 5-6 m long and 4 m wide, at most 0.5 m deep, with 0.5 l/sec of 45°C water.

Dynjandislaug

GPS N65 44.079-W23 12.439 WGS 84

30 meters asl

Dynjandislaug is in the Dynjandisvogur inlet off Arnarfjörður. To reach it, turn off road no. 60 to the travellers' facilities by Dynjandis-foss waterfall and drive down to the campsite south of Bæjarfoss falls in the river Dynjandisá. Beyond the campsite you can just see a track leading up the grassy slope to the ruins of a farm on the brow of a little hill. Beyond the ruins is a marshy area, created by the runoff of the warm stream from the pool. You cross this marsh to get to the pool, and the best way to avoid the worst of the wet is to swerve round it towards the river Dynjandisá.

Þóra Sigurbjörnsdóttir checks the water in Dynjandislaug. Photo: Jón G. Snæland

Ruins below the pool. In the distance is Borgarfjörður, where the Mjólká power station is. The mountain on the right is Meðalnesfjall. Photo: Jón G. Snæland

The pool is well-built of turf and stone. It is 1.2 x 2 m, about 30 cm deep and the temperature is 25-30°C. There is a lot of algae in the pool, making it rather murky. There are no facilities by the pool, but clothes can be laid on the grassy slope.

Dynjandisfoss waterfall in Arnarfjörður. Photo: Jón G. Snæland

The Dynjandisvogur inlet, overlooking the campsite beside Dynjandisá.
Photo: Jón G. Snæland

The river Dynjandisá runs down off Dynjandisheiði heath, south
of the mountain Gláma, which was a glacier until 1950, and still has
large year-round snowdrifts. Dynjandisvogur inlet is famous for the six
waterfalls in its river. The top one, most famous of all, is Dynjandis-
foss. It is 100 m high, 30 m broad at the top of the falls, broaden-
ing to 60 m at the bottom. The other five waterfalls are: Hundafoss,
Göngumannafoss – which you can walk behind – Hrísvaðafoss,
Sjóarafoss and Bæjarfoss, north of the campsite.

The farm Dynjandi was situated west of the river, but it was abandoned long ago, and there is little to remind one of it.

Dynjandisfoss and its surroundings, including the thermal pool,
were designated a protected conservation area in 1981

Galtahryggjarlaug

GPS N65 50.396-W22 40.676 WGS 84

57 meters asl

This pool is in the valley Heydalur on the west of Mjóifjörður in Ísa-fjarðardjúp, near the bottom of the fjord. You drive in the valley to the farm Heydalur, where you should speak to the inhabitants about visiting the pool, since the track lies right past their front door. After that the track leads below the farm buildings, through a gate and across the Heydalsá river. Once across, it is best to leave the vehicle on the bank, since there is a marsh between the river and the pool. The river bank is in fact where the GPS was taken, about 20 m below the pool itself.

The pool is about 2 x 4 m and 50 cm deep, the water temperature around 40°C and the flow 0.2-0.3 l/sec. It is built of stone and turf with a fine gravel bottom so it stays clear, and the surroundings are grassy. There is a 3 m² changing shed with a bench and clothes hooks

Galtahryggjarlaug is in Heydalur, and is often called Heydalslaug. Photo: Jón G. Snæland

Enjoying the calm of morning in the natural themal pool in Heydalur. Photo: Siv Friðleifsdóttir

is a stones throw from the pool. The area around the pool is a registered natural monument and the ruins of the abandoned farm Galtahryggur a registered archeological site.

Hörgshlíðarlaug

GPS N65 49.861-W22 37.733
2–3 meters asl

Hörgshlíðarlaug on the east of Mjóifjörður in Ísafjarðardjúp, near the bottom of the fjord, is approached along road no. 633. The pool is below the road, about 2 km south of the farm Hörgshlíð to which it belongs.

The pool is concrete, 2 x 6 m and about 0.8 m deep. Two hot water hoses and one cold provide water with a temperature of about 40°C flowing at 1 l/sec. There is a wooden changing shed at the end of the pool with a bench and clothes hooks.

Finnbogi Jónsson and his family at Hörgshlíð own the pool, and everyone is welcome to bathe there, so long as they treat it with respect and get permission from the owners if they are at home. There is a collection box in the shed where those who care to can make a contribution towards the expenses of the upkeep.

A bridge crossing Mjóifjörður is under construction, which will seriously reduce the traffic visiting the inner end of the fjord.

The Hörgshlíð pool is clean and well-maintained, a credit to its owners. Photo: Magni Rúnar Þorvaldsson

Hörgshlíðarfjall

GPS N65 51.412-W22 34.750
127 meters asl

The track to the mountain Hörgshlíðarfjall starts at the farm Hörgshlíð in Mjóifjörður. This is a steep, 2.4 km jeep track up along the mountain. Once you are up, a clearer track leads another 1.5 km to the lake Fremra-Selsvatn. The main track leads right to the valley Hörgshlíðardalur and an impressive lodge, moved up there by some jeep enthusiasts from a golf course in Ísafjörður. Just beyond the lodge is a small building containing a changing room with a small covered pool outside it. The pool, which is protected by a wooden lid, seats 4-6 on underwater benches.

Members of a jeep club and a flying club examining the thermal pool on Hörgshlíðarfjall. Photo: unknown

Keldulaug

GPS N65 53.424-W22 34.980 WGS 84
30 meters asl

Keldulaug in Mjóifjörður in Ísafjarðardjúp is on the east side of the fjord below Kelduhæð ridge south of the farm Keldur. As you drive out along the fjord on road 633 you come to a little stream with light-coloured mineral deposits above the road (GPS N65 53.430-W22 35.056). From there you can reach the pool by either walking 100 m up the hill, or driving a little further along the 633 and turning in towards a summer house, through a gate and along a track (GPS N65 53.648-W22 34.795). The pool is privately owned, by Ari Sigurjónsson of Ísafjörður.

The concrete pool is 2 x 2 m, about 90 cm deep with a 1 l/sec flow of 40°C water. It is well protected from the wind and attached to a changing room with two benches and some clothes hooks. The pool was built about 20 years ago is kept very clean and neat, as are its surroundings.

Keldulaug is one of those pools that are very well kept up by their owners. Photo: Jón G. Snæland

Reykjanes

GPS N65 55.651-W22 25.657 WGS 84

12 meters asl

The Reykjanes thermal pool was built around 1925-27, and a boarding school was built beside it in 1934. All teaching, including swimming teaching, stopped there about 50 years later.

The pool is concrete, 52 x 12.5 m, and from 0.6 to 2.3 m deep. The temperature is from 36-40°C, the northern end being the cooler. The water comes from a borehole and the run-off from the hotel's (formerly the school's) heating system. There is no chlorine or other chemicals in the water.

There are changing rooms for 20 men and 20 women in the hotel, and five showers. Out by the pool is a steam bath with room for 8. The steam comes from the geothermal water.

For several years tomatoes and other vegetables were grown commercially in greenhouses beside the pool. Right below the hotel is a pier and a bit further in the fjord salt production was started in 1773.

The school in Reykjanes where thousands of pupils have boarded over the years. It is now a hotel. Photo: Jón G. Snæland

Swimmers in Reykjaneslaug. Photo: Valborg Jónsdóttir

It produced good salt, but it could not produce enough to be viable so production lasted only 13 years. On the seashore in Hveravík are some boiling springs which go underwater at each high tide.

Reykjanes (old pool)

GPS N65 55.384-W22 25.417 WGS 84

20 meters asl

To reach the old pool at Reykjanes in Ísafjarðardjúp drive about 900 m in the fjord from the Reykjanes school (hotel). When you reach a little hill (GPS N65 55.428-W22 25.339) you should see a sign by the road. There you will find steps leading to a 100 m walk from the sign to the pool.

Though the pool has become quite run-down, with little water (but plenty of stickleback and wading birds) you can see that it was a notable construction in its day. Water runs into it from the nearby Kolahver hot spring. The pool was built of turf by the local authority in 1889, and later enlarged and improved using concrete. They taught swimming in this pool for 97 years, from 1830 to 1927.

The run-down old pool in Reykjanes. Langaströnd on the other side of Ísafjarðardjúp in the distance. Photo: Jón G. Snæland

Gjörvidalslaug (Gervidalslaug)

GPS N65 47.687-W22 31.627 WGS 84

20 meters asl

On the east side of the fjord Ísafjörður is Gjörvidalslaug, just above road no. 61. The GPS coordinates given were taken on the highway, but the pool itself is about 50 m above that.

There are two Gjörvidalslaug pools, the main concrete one and a small pool a few meters higher up. The concrete pool is 2 x 1 m, 60 cm deep with 42°C water flowing at 0.2-0.3 l/sec. The couple Dúna and Jens in Gjörvidalur look after the pools, which are government property. They have built a nice shed by the pool with a little deck where there is a bench and some clothes hooks.

Above the main pool is a little stone-built pool at a temperature of 44°C where it is fun to soak your feet. There is a third pool on

The small pool above the main pool. Photo: Jón G. Snæland

Gjörvidalslaug was popular with travellers in the days when the trip between Ísafjörður and Reykjavík took longer than it does today. Then it was just the place to rest weary limbs along the way. Photos. unknown

Gjörvidalur land, called Þvottalaug (Washing Pool), but it is kept for the private use of the residents.

Nauteyrarlaug

GPS N65 55.027-W22 20.516 WGS 84

35 meters asl

This pool is on the Nauteyri farm on Langadalsströnd deep in Ísafjarðardjúp, 200-300 m beyond the Nauteyrarkirkja church. You turn off road no. 653 at GPS N65 54.960-W22 20.628 from where you drive 100-200 m to reach the pool. There is a one-room house by the pool for changing clothes.

Nauteyrarlaug is an old, stone-built sunken bathing pool which has room for lots of people and does not easily get cloudy. The water that runs into the pool through a hose is rather hot, over 40°C, but can be cooled by letting the water run through the grass for a bit. The water running into the pool is about 42-43°C, but 35°C running out. The acidity of this water has been measured at pH 9.7.

Nearby is the church Nauteyrarkirkja, built in 1885 of wood with corrugated iron cladding. There is seating for well over 100 in the church, which has a number of remarkable possessions, including a chalice from 1750, made by Sigurður Þorsteinsson, a Copenhagen silversmith.

A group of volunteers bathing in Nauteyrarlaug. Photo: Jón G. Snæland

Nauteyrarpottur

GPS N65 55.873-W22 22.458 WGS 84

11 meters asl

Below road no. 635 leading to Snæfjallaströnd there is a salmon fishery, about 300 m south of which is the thermal pool called Nauteyrarpottur. It is about 1.5 x 1.5 with 43°C water piped into it. The surroundings have been designed to prevent water from collecting around the pool, so clothes can be laid in the grass. Brooms are kept there for cleaning the pool, or skimming dead flies off the top. When last this writer visited the pool there was a fair collection of insects in it.

To get to the pool drive out road 635, past the Nauteyrarkirkja church and turn off down to the road to the salmon fishery. Take the first turn to the left onto an old track. About 20-30 m below the track is the pool, against a grassy bank.

Nauteyrarpottur with the salmon fishery and the Langadalsströnd coast in the background. Photo: Jón G. Snæland

Laugarás

GPS N66 01.178-W22 23.231 WGS 84
23 meters asl

The farm Laugarás is in the valley Skjaldfannardalur in Ísafjarðardjúp. Above the farm Skjaldfönn is a mound-shaped snowdrift which seldom melts completely over the summer, giving the valley its name ("Shield snow valley"). It used to provide an easy route onto the Drangajökull glacier, where wood was often carried across by horses. Right next to the farmhouse there is a nice concrete private pool, built in 1985. The pool is 5 x 10 m, 0.5 to 1.25 m deep with a flow of 1.3 l/sec. There are no changing facilities.

The farm Laugaland, a short distance further in, is usually found on maps, but usually not Laugarás. Both are on the south-west side of the valley, opposite the farm Skjaldfönn. Presently, the farmer at Laugarás is Þórður Halldórsson. Those who want a swim should consult Þórður or the folks at Laugaland.

Laugaráslaug and the former family home. Photo: Jón G. Snæland

Hestvallalaug

GPS N66 15.282-W22 05.178

15 meters asl

There are numerous hot springs in Reykjarfjörður, a short, open fjord between Þaralátursfjörður and the headland Geirólfsgnúpur at the southern end of the Hornstrandir coast. The first farm was established in 1876, but there has been no farming there since 1964.

In 1937 the sons of the farmer built a concrete pool and piped in water from the hot spring, one of the sons proceeding to teach many of his neighbours how to swim. The pool is only in use in the summer and has been well kept up and improved over the years.

Members of the family come every summer to collect driftwood which they saw into fenceposts in their workshop at Hlein.

The swimming pool in Reykjarfjörður. Photo: Sigríður Lóa Jónsdóttir

Krossneslaug

GPS N66 03.357-W21 30.400
On the beach

Krossneslaug is by the Laugavík inlet. To get there, drive from the village of Norðurfjörður, past the farm Krossnes out onto the peninsula, then to the north. This will bring you to the first of two small parking areas, 50-100 m apart. Below the road between them is Krossneslaug, about 10 m from the water at high tide. There is an extremely steep track down, ending in a very rough gravel bank too dangerous for single drive vehicles, which should be left in the parking areas.

The pool is concrete, about 6 x 12 m and 0.6-1.8 m deep, with a temperature of about 31°C, though the hot springs themselves can reach 64°C. There are good changing and showering facilities for several dozen people, and the pool can take 30. Access to the pool is open, but adults are expected to pay ISK 200 each. The Youth Association *Leif the Lucky* owns the pool, which was built in the early 1950s.

Krossneslaug on the beach in Laugavík inlet. Photo: Jón G. Snæland

On the beach near Krossnes is a stack called the Thirty-Dollar Stack. The story goes that a boat captain once climbed it, left 30 dollars and then said that anyone who could climb the stack could claim the dollars. Near the end of 1899 a fisherman from Eyjafjörður managed the climb, but found no coins.

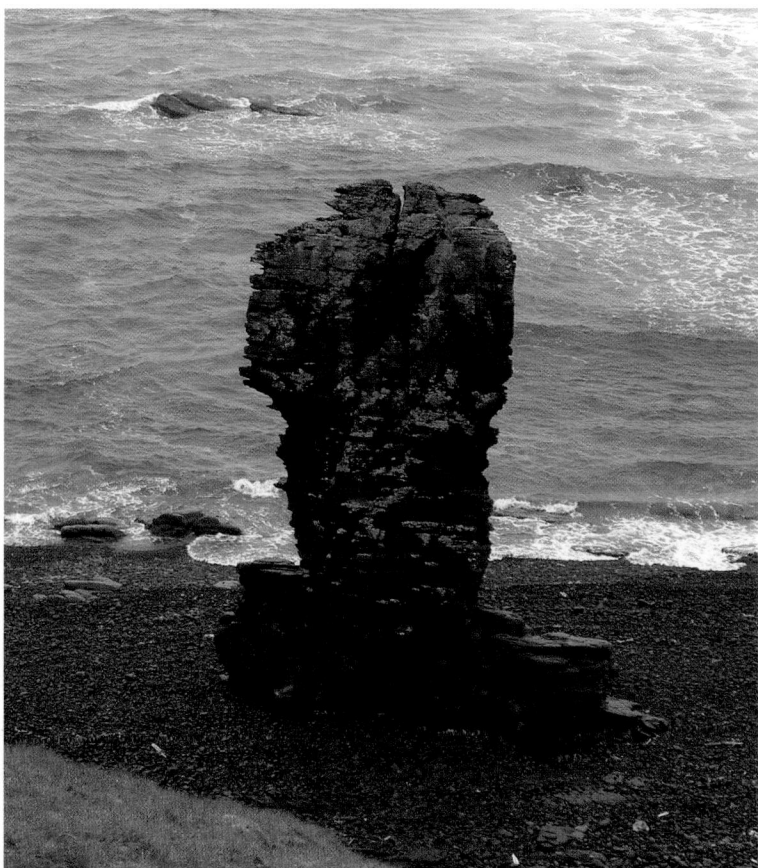

It was said that thirty coins had been left on top of Þrjátíudalastapi, to belong to whoever could climb it.

Hákarlavogur

GPS N65 59.931-W21 19.021
2 meters asl

Gjögur has long been a famous fishing station. Not far out to sea were rich shark fishing grounds. From the 19th century to 1930 there was so much shark fishing from Gjögur that as many as 30 shark boats were stationed there at once.

Hákarlavogur is just beyond Gjögur. Drive along the airstrip, past the terminal and around the end of the airstrip. There the road leads down across a grassy area down towards the rocky crags in Akurvík off Hákarlavogur inlet. A short way onto the crags is a natural pool, one of Icleand's finest, as though hewn into the cliff. Nothing is man-made except a small concrete dam across a gap in the rocks. There is room for 6-10 bathers and the hot water running into it is 65°C. The pool and its surroundings are the property of a branch of the Thorarensen family, who look after it. It is not open to the public.

The pool in Hákarlavogur. Photo: Soffía Snæland

Gvendarlaug

GPS N65 46.867-W21 31.148
10–15 meters asl

The pool is at Laugarhóll (Klúka) in Bjarnarfjörður. The name comes from the fact that Guðmundur the Good Arason, bishop at Hólar, consecrated it as a bathing pool in 1237. Gvendarlaug is a neatly made, circular, stone-built pool about 2 m across. The walls are vertical but there is a lot of algae in the water so it is difficult to judge the depth. The pool was renovated not so long ago and has been under the protection of the National Museum since 1989. Bathing is not allowed in this pool, which is half way between the hotel and the changing rooms by the swimming pool. Historical information is provided by the pool.

Down by the swimming pool, by a grassy bank, is another natural pool which has also been called Gvendarlaug. It is sandy-bottomed, the temperature about 42°C and there is room for 2–3 bathers. The changing rooms by the main pool are available to all bathers.

The main pool is called Gvendarlaug the good. It was built in 1943-1947 by the Grettir swimming club, and is a fully equipped

Gvendarlaug. Photo: Jón Þorsteinsson

The main pool at Klúka. A smaller pool can be seen below the main pool, and there is yet another pool under the bank where the walkway ends. Photo: Jón Ebbi Halldórsson

outdoor swimming pool. The pool is open daily from 10–22, but access can be obtained at other times from the hotel staff, when the hotel is open.

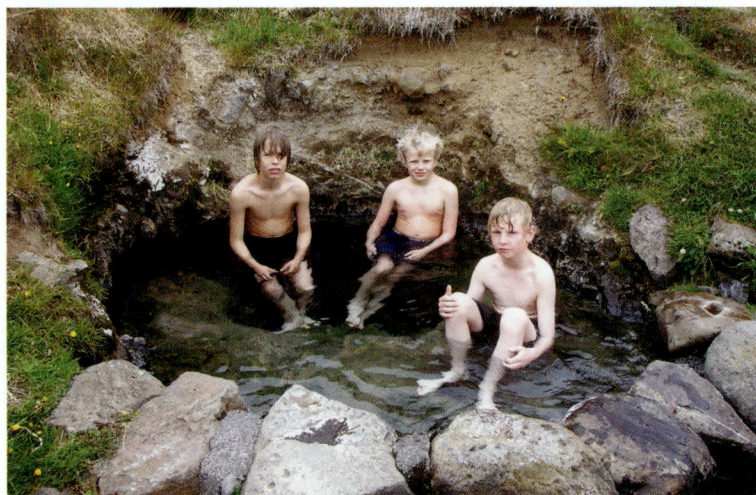

Three friends bathing in one of the Gvendarlaugs. Photo: Jón Ebbi Halldórsson

Drangsnes

GPS N65 41.296-W21 26.894 WGS 84
On the foreshore

Alongside the breakwater on the foreshore on the main road in the village of Drangsnes are three hot tubs, two standard and one fishing tub. They are side by side on a neat, specially built deck. The water comes from a nearby borehole and is about 40°C.

They found hot water in Drangsnes in 1997, which was a great advantage for the village. Within days of the find, the locals had set up the bathing area. The hot tubs are maintained by the local authority and are open to all.

Admiring the hot tubs in the middle of the village of Drangsnes. Photo: Þóra Sigurbjörnsdóttir

The North

Hveraborg

GPS N64 59.906-W20 55.905 WGS 84

345 meters asl

After crossing the river Hrútafjarðará by the old bridge, turn right onto a track marked Hveraborg, east of the river Síká. Though marked as a jeep track, it should, with care, be passable by all vehicles. It is a good gravel road as far as the abandoned farm Fosshóll. After that you have to ford the shallow river Brennikvísl and a few streams. At the end of the 11.5 km road there is a fenced-off area, which is a convenient place to leave your vehicle and start your 6.9 km walk to the pool. After going over the fence, walk first along the continuation of the track, and then along clear, much-used sheep trails which still have a few trail markers on them, though many have fallen. The trail more or less follows the course of the Síká and takes 1.5 to 2 hours each way. The ascent is only about 100 m. Hveraborg is in an area of the high-

The pool in the river Síká is sometimes damaged in the spring floods and in need of repair. In this photo they have tried to improve it using industrial plastic. Photo: Jón G. Snæland

The North

Pools are marked with a star.
Pools marked with a green star
are not specifically covered.

Sigluf

Ólafsfjörðu

Skagafjörður

⭐ Grettislaug

hAflói

Sauðárkrókur

⭐ Biskupalaug Hj

Blönduós

⭐ Vallarlaug

⭐ Reykjavellir

⭐ Hveraborg

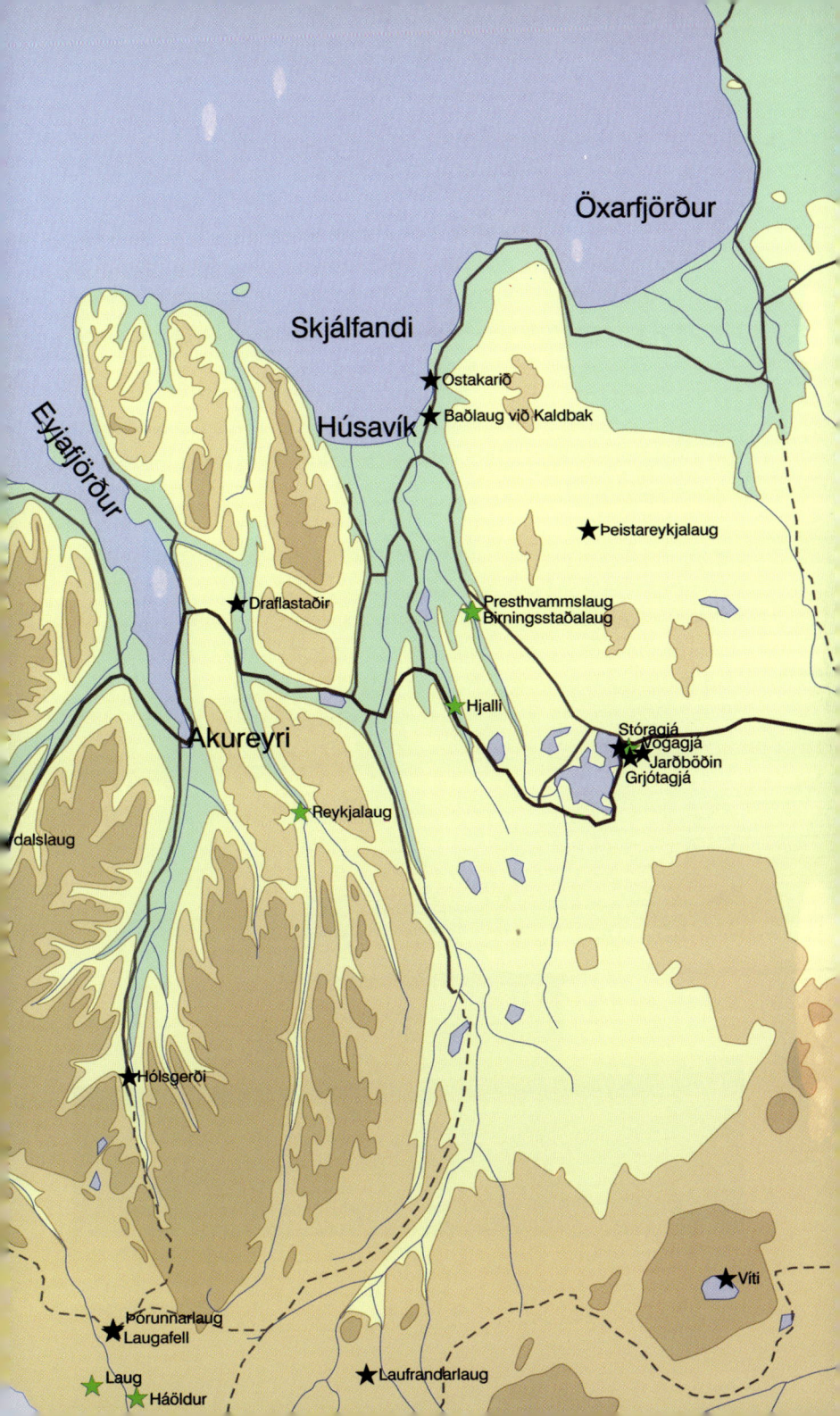

Öxarfjörður

Skjálfandi

★ Ostakarið

Húsavík ★ Baðlaug við Kaldbak

Eyjafjörður

★ Þeistareykjalaug

★ Draflastaðir

★ Presthvammslaug
★ Birningsstaðalaug

★ Hjalli

Akureyri

Stóragjá
★ Vogagjá
★ Jarðböðin
Grjótagjá

★ Reykjalaug

dalslaug

★ Hólsgerði

★ Víti

★ Þórunnarlaug
★ Laugafell

★ Laug
★ Háöldur

★ Laufrandarlaug

lands called Tvídægra, meaning "a two-day journey". This is because it used to take two days to cross the highlands from Húnavatnssýsla County to the south.

On the western slopes of the mountain Sléttafell is Sléttafellshverir, or Hveraborg, a sizeable area of geothermal springs coming up on both the banks and the bottom of the Síká. The thermal pool is on the eastern side of the river, below the Hverarborgarskáli lodge. The pool is rock-built, sealed with concrete and plastic foam. Since the pool was built, the river Síká has not always been kind to it, so one side has collapsed. People have tried with limited success to rebuild it and seal it with plastic sheeting. The water used to be 60 cm deep, but is now between 20 and 30 cm.

The pool's size is 2 x 2 m and the bottom is smooth rock. Water is piped into it from a small hot spring above the pool. The water in the hot spring is said to be 69°C, running at 0.5 l/sec, and the temperature when it reaches the pool 37°C. When this writer was there in 2008 the water seemed hotter, so it had to be cooled with water from the river.

A little way above the pool is the lodge Hveraborgarskáli (GPS N64 59.922-W20 55.844 WGS 84). It is the property of the Vestur-Húna-

The Hveraborgarskáli is on the banks of the river Síká, just above the thermal pool. Photo: Jón G. Snæland

vatnssýsla Travel Association and maintained by the staff at Staðarskáli, who keep the key. The lodge was built 1998-99, but is already in some disrepair. It is poorly equipped and hardly usable except on a warm summer's day. There is room for 10-12 on mattresses, and there are chairs and tables.

The pool nearer the beach was full of stones after high seas. Photo: Jón G. Snæland

toilets and tables and benches. There are plans to open a coffee shop in another turf house, offering home baking such as hot waffles. They have built their own 3 kw mini-hydroelectric station in a stream on the hillside.

One of the turf houses by Grettislaug, the washroom and WC. Photo: Þóra Sigurbjörnsdóttir

Biskupalaug

GPS N65 39.625-W19 04.759 WGS 84

220 meters asl

Biskupalaug is deep in the Hjaltadalur valley. Take road no. 767 and then 768. When you reach the turn off to Reykir go left across a stream and drive 830 m to a gate. Drive through the gate to a borehole, leave your vehicle and walk past the borehole and over a small stream. Once across the stream the pool is a few dozen meters further on, slightly to the right. It lies in a small dell by a grassy bank and can be quite difficult to find, there being no paths or other signs to lead you to it.

The pool is oval, with stone edges, 2.6 x 3.5 m with a depth of up to 0.8 m. Since they bored for hot water nearby, the pool has lost some heat, and is now just barely warm. There is a lot of algae floating in the water, and vegetation rotting on the gravel bottom, so the water quickly gets dirty with use. The water streams up through the gravel bottom. Though there are no facilities nor anything else man

Biskupalaug in the Hjaltadalur valley is a protected archeological site. Photo: Jón G. Snæland

Biskupalaug is not easily seen until you are standing on its banks. In the distance is the borehole that supplies water to Hólar. Photo: Jón G. Snæland

made near the pool, the grass is thick around it, fine for laying your clothes on. They say the bishops at Hólar used to bathe here, about 10 km from Hólar. Since 1969 the pool has been designated a protected archeological site. There used to be another, larger pool in the area called The Labourers' Pool.

Hörgárdalslaug

GPS N65 34.527-W18 42.165

376 meters asl

There is a thermal pool deep in the Hörgárdalur valley in Eyjafjörður. From Highway 1, the ring road, in Öxnadalur turn off onto road no. 814. It leads into Hörgárdalur and past the driveway to Staðarbakki, the innermost farm. From there, 5.4 km along a jeep track will bring you to the pool. When this writer last drove there in 2007 it seemed to be passable by jeeps with at least 33 inch wheels. The track forks, one fork leading straight ahead to a ford in the Hörgá river. The track to the pool, on the other hand, goes up along the hillside by the abandoned farm Flögusel, after which the track gets worse. From there it goes over a ridge and down to gravel banks on the Hörgá, called Laugareyrar, and this is where the pool is.

The pool can be difficult to find without GPS. It is about 5 m from the track and 2-3 m from the river. It is about 3 m long, 1.5-2 m wide

Daníel Hartmannsson checking out Hörgárdalslaug. Photo: Jón G. Snæland

and 40 cm deep. The water temperature is about 30°C and the inflow is not great. It was once in much better shape, with stone-built walls and thus deeper, but in recent years it has been almost completely destroyed by the river when in spate.

Further in the valley there is a cairn-marked walking route over to Hjaltadalur. The cairns are closer together once you reach Hjaltadalsheiði. Travellers who would like to extend their day's journey by jeep can drive back out Hörgárdalur to where the mountains on the left open out, just before the confluence of the rivers Hörgá and Barká. Right by a summerhouse there, a track leads into Barkárdalur valley, with the Barká river running immediately to the right of the track. The track makes its way along gravel moors and then along the river gorge below Slembimúli ridge. Across the valley you can see the abandoned farm Féeggsstaðir and the river Féeggsstaðaá tumbling in waterfalls down to Barká. Below and to the south of the farmhouse is a sheep's (foot) bridge over the majestic Barká gorge. A little further south there is another bridge – too narrow and weak for modern jeeps – and a ford. The river is rather deep at the ford, 50-60 cm in dry weather and can increase without warning in wet weather, so it is dangerous for all but the largest jeeps. From the ford you can see an old, abandoned turf farmhouse, Baugasel, which has been renovated as a travellers' hut.

At the end of the Barkárdalur valley you come to some old walking tracks between communities. The Hólamannavegur track is probably the highest road in Iceland, at 1200 m asl. The track is for walking only, not even passable on horseback. Another walking route is Tungnahryggjarleið. Near the glacier Tungnahryggjarjökull there is a roundup hut.

Hólsgerðislaug

GPS N65 18.428-W18 15.300 WGS 84

181 meters asl

Hólsgerðislaug is deep in the valleys of Eyjafjörður. Drive south on road no. 821 along the west side of the fjord all the way to the end, at the farm Torfufell, and then along mountain road F821.

Drive to GPS N65 18.419-W18 15.557 which is on the road directly above the pool. There is a summer house to the north of the pool. Leaving the road you have to cross an electric fence, then walk about 200 m in a straight line down from the road across a grassy slope. The pool, which is stone-built in part, is under a high, grassy bank. It is 4.5 x 2 m, 50-70 cm deep with a temperature of about 42°C and room

A group of travellers from the Útivist Travel Association at Hólsgerðislaug. Photo: Jón G. Snæland

Dagur Bragason by the St. Peter's cairn on Vatnahjalli. Photo: Jóhann Helga-son

for 6-8 bathers. The bottom is gravel and some large rocks. The inflow is slow, and it does not get cloudy with use. The owner of the pool is Brynjar Skúlason of Akureyri.

Hólsgerðislaug is at the start of the highlands, the road above the pool running south to Sprengisandur.

Draflastaðir

GPS N65 49.301-W17 54.522 WGS 84
87 meters asl

The farm Draflastaðir is on the west side of the Fnjóskadalur valley.
Take road no. 834 off Highway 1, where it comes down from Víkur-
skarð pass. Drive 4.6 km to the driveway to Draflastaðir, pass the
stables there and drive another 887 m along a track passable by all
vehicles. The track is rather narrow for parking, but there is very little
traffic. From the road, climb over a barbed wire fence, cross a deep
ditch and continue another 100 m to the pool. You will you see an
upturned plastic tub that covers the borehole machinery. The pool
itself is a kind of pond into which the borehole run-off is piped.

The pool is about 6 x 9 m, 30-40 cm deep water temperature about
33°C. The flow is 0.7 l/sec. There is a lot of mud on the bottom and the
pool gets so cloudy as to seem black. It is not tempting to bathe in it,
but it could be improved if it was deepened and the mud dug out of it.
There should be room for up to 30 bathers. There are no facilities and

Draflastaðalaug. Photo: Jón G. Snæland

the edges to increase the heat of the water at the hose end of the pool, creating two pools of different temperatures.

The tub is used a lot by the residents of Húsavík, not least by those who suffer from psoriasis or eczema. A 20-foot container provides an electrically lit changing room, with hooks but niether shower nor WC. It is not divided into men's and women's changing rooms.

The town of Húsavík owns the pool, but it is maintained by volunteers.

Pond by Kaldbakur

GPS N66 00.953-W17 21.556
About 50 meters asl

Just to the south of the town of Húsavík, above the highway by Kald-bakur, is a pond that has been dammed. 20-30°C hot water is piped into it from the Húsavík Energy power station. This is just hot enough so that the children of Húsavík tend to frequent it in the summer. The surroundings are all very attractive. To approach the pool you cross a pretty arched bridge over the run-off stream and there is lots of gree-nery all around the pond.

A newspaper (Fréttablaðið) story in January 2006 told of a huge goldfish (34 cm long, weighing 850 gr) that was found there. App-arently some unknown person had released 5-6 ordinary pet goldfish into the pond a year earlier. They were only a few centimeters in length then, so the pond would appear to provide an excellent habitat.

The pond is surrounded by all kinds of greenery, so its shore is perfect for a picnic. Photo: Jón G. Snæland

Þeistareykjalaug

GPS N65 52.781-W16 57.268

353 meters asl

The pool is below the mountain Bæjarfjall, to the west of Þeistareykja-bunga about 150 m east of the Þeistareykjaskáli lodges. It is only a few years old, made by using turf and stone to dam a warm stream that carries the run-off from a borehole about 100 m away.

The water is a good temperature for bathing and about 1m deep, with room for lots of bathers at the same time. The bottom is clay so it gets cloudy with use. Clothes can be laid on the grass. The nearby Þeistareykjaskáli lodges are run by the Húsavík 4x4 club and Húsavík snowmobilers. The more recently built lodge is from 1958, well equipped and kept up, and open to travellers. There is room for 30 people, 18 in bunks. It is geothermally heated, has a gas cooker and running water.

Þeistareykjalaug with the lodges in the distance. Photo: Dagur Bragason

There are many sights in the area, for instance the cave Togarahellir which is on the direct line between Mt. Mælifell and Ketill in Mt. Ketilsfjall, a walk of 30-40 minutes over rough lava. In the book *Lava Caves in Iceland* its situation is described as follows:

Togarahellir is at the end of sizeable lava channels about 3 km north-west of the abandoned farm Þeistareykir on the west of Þeistareykja-bunga. Size of the cave: length about 100 m, width about 15 m and height 10-12 m.

Those who have a jeep, or are interested in a longer walk, can make their way up above the lodges and through Bóndhólsskarð pass and east to the Víti lakes Litla-Víti and Stóra Víti, about 5 km as the crow flies.

Stóragjá

GPS N 65 38.304-W16 54.591
282 meters asl

As you turn east from Lake Mývatn through the village of Reykjahlíð on Highway 1, the well-marked path to Stóragjá is south of the road. You can see into the gorge from the path, but it is about 170 m to the stairs down into it. The stairs are good, fairly new, 35 steps in two sections. Once down you are in a sort of wonderland where you walk a few dozen meters along the canyon on an uneven path with the boulders towering above you. The water is often visible between the rocks and you can either let yourself down by a rope into the clear, 29°C water, or reach it down more steps at the southern end of the gorge. The depth varies wildly since the water is full of boulders and large rocks. There are no changing facilities, but plenty of rocks to lay your clothes on. There is, however, a serious problem with this heavenly pool: e-coli bacteria have been found in it, so it is no longer advisable to bathe there.

Stóragjá. Photo: Þóra Sigurbjörnsdóttir

Because the sides of the gorge are pretty high, a staircase has been built leading down into it. Photo: Jón G. Snæland

Grjótagjá

GPS N65 37.593-W16 52.974
289 meters asl

Grjótagjá is in a lava fissure about halfway along the short road no. 860, that runs between two points on Higway 1, the tourist services at Vogar and the road a little east of the village Reykjahlíð. This gravel road is fine for all vehicles and well maintained. There is a parking area just by Grjótagjá.

There are two entrances, one for either sex, with a section of the pool for each. From the cave mouth you clamber down 2-3 m to the surface of the water, which is 2-3 m deep ground water. The pool is 7-8 m long and 3-4 m wide. The temperature of the water rose sharply during the volcanic activity of 1975-84. Measured in 2004 it was still 47°C but it is cooling steadily and should soon reach bathing temperature again. There is room for 20-30 bathers, who can leave their

The cave mouth of Grjótagjá. Photo: Þóra Sigurbjörnsdóttir

Helena Snæland checks out Grjótagjá, but the water is still rather hot after volcanic activity. Photo: Þóra Sigurbjörnsdóttir

clothes on a rock. The fissure is kind of a cave, nearly closed above by old lava. The roof is a collection of huge rocks, some of which look as if they are about to fall into the water. Many travellers just stick their heads in and having seen the situation decide they want nothing more to do with it. Grjótagjá belongs to the farm Vogar.

Jarðböðin við Mývatn (Mývatn Nature Baths)

GPS N65 37.861-W16 50.864

334 meters asl

Jarðböðin við Mývatn were formally opened in June 2004, but the history of bathing in natural hot water and mud at Mývatn is a long one, probably reaching back to the settlement years. Old books tell how Bishop Guðmudur the Good consecrated such a bath in the 13th century.

Through the ages various huts have been raised over and around places where there is steam or hot water. Sources from the 16th, 17th and 18th centuries describe them, usually with positive comments about their beneficial effects, though occasionally the writer has found the heat uncomfortable, and an ancestress of the Skútustaðir family is said to have suffocated in a steam bath in 1803.

In the 20th century bathhouses were built in the 1940s and 50s, and were in use until 1970. In the 1990s they built facilities yet again

The Mývatn Nature Baths have been built up and modernised in recent years, so now the facilities are excellent. Photo: Jón G. Snæland

The Mývatn Nature Baths. Photo: Jón G. Snæland

which were eventually enlarged and improved, ending in the present bath house with its large pool and full facilites, including a restaurant.

The Nature Baths, with their mineral rich water, have become very popular, with thousands visiting them each year.

The East

The East

Pools are marked with a star.
Pools marked with a green star
are not specifically covered.

Þistilfjörður

Bakkaflói

Vopnafjörður

Héraðsflói

Seyðisfjörður

Egilsstaðir

Norðfjörð

Reyðarfjörður

★ Laugavallalaug
★ Laugarhús

★ Lindur

★ Laugarfell

Breiðdalsvík

Berufjörður

Laugarhús

GPS N64 59.252-W15 35.579
443 meters asl

The ruins of Laugarhús farm are deep on the east side of the valley Hrafnkelsdalur, which leads south from the Jökuldalir valleys. The farm was named for the thermal pool or spring which comes up above the farmhouse and now sends a stream down through the ruins. Laugarhús has a long history and at some point a pool was built there, with a stone bottom. In the 1397 records of the church at Valþjófsstaður, the church is said to have a mountain dairy at Laugarhús. Signs of human habitation are found in a large mound, about 20 m across. They have drilled through the mound here and there and examined the cores to discover their age. What they have found is that repeated building went on there in medieval times, but most of the ruins appeared to be from the 14th century.

There is not much left of the Laugarhús pool, but it could be greatly improved without much trouble. Photo: Jón G. Snæland

The Hrafnkelsdalur valley takes its name from Hrafnkell Freysgoði, the hero of Hrafnkels Saga, who lived at Aðalból. The Saga says that Hrafnkell's neighbours and long-time enemies, Bjarni and his sons Sámur and Eyvindur, lived next door at Laugarhús.

To reach Laugarhús enter Hrafnkelsdalur in the north, crossing the river Jökulsá á Brú by the farm Brú. From there the most obvious, but not always passable, route is to fork left and drive south along the valley. Eventually you reach Aðalból, where there are various services for the traveller; a little shop, a petrol station, indoor accommodation or camping in the home field. A little further on you have to ford the river Hrafnkelsá. However, it is often too deep for smaller jeeps to ford safely – this was for instance the case when this writer was there in the summer of 2008.

Another route via Brú is to head for the southern end of Hrafnkels-dalur. Fork right and drive up out of the valley, past the farm Vað-brekka and south along the Vaðbrekkuháls ridge to road no. 910. Turn left and drive approximately 15 km til you reach the turnoff onto F910 on your left, and drive into Hrafnkelsdalur on the east. Alternatively approach the F910 from the direction of Egilsstaðir along road no. 910 and take a right.

When you reach the valley floor you are virtually in the Laugarhús farmyard. Park by the sheepcotes. As you walk up, you should cross a fairly deep stream at the corner of the sheepcotes where a narrow green plastic hose lies across it. Once on the bank of this stream you should see a small stream which you follow up the way for a few dozen meters. There you will find a warm spring which was at one time a pool. Little or nothing remains of the former pool, which has clearly been washed away by streams in spring spate. All that is left is the spring, which is a good bathing temperature, so it should be easy to build a new pool.

Laugarfellslaug

GPS N64 53.138-W15 21.140
550 meters asl

Laugarfellslaug is northeast of the mountain Snæfell. Drive south along road no. 910 on Fljótsdalsheiði heath. Towards the southern end of the heath you are nearing Laugafell. There you turn east on a side road that goes down along the river Laugará, by a road sign incorrectly marked *Laugafell 4 km*. Somehow this sign has come there from a crossroads on Sprengisandur, where it pointed to Laugafell-without-an-R. From this turnoff it is 2.4 km to the Laugarfellsskáli hut. All vehicles can drive down to Laugará which flows there through a lovely gorge with a great many beautiful waterfalls. The best-known is Slæðufoss ("Scarf Falls"), above the ford, and further down is the lovely Stuðlafoss. The river can only be crossed by jeep since the ford is very stony and rough. It is not far from the ford to the Laugarfells-skáli hut. There even jeeps must be left. Walk to the pool which is near

Above Laugarfellslaug is a cattle shed which can be used as a changing room.
Photo: Þóra Sigurbjörnsdóttir

Laugarfellslaug and the herders' hut. Photo: Jón G. Snæland

a cattle shed which can be used as a changing room. It is also fine to leave your clothes on the grass by the pool, which is dug into an earthy soil and lined with large, flat stones. The pool is circular, around 3.5 m in diameter and 70 cm at its deepest. The water flows in through the bottom and there is a run-off channel to keep the surface even. The water temperature is around 34°C and at least a dozen people could bathe at the same time. The pool is registered to the vicarage at Val-þjófsstaður.

Laugarvalladalur

GPS N65 00.390-W15 45.734
535 meters asl

The pool is in the Brúardalir area north of the dam and Kárahnjúkar.

Only a few years ago not many had heard of the Laugarvalladalur valley, other than local or mountain people. The valley was difficult to get into, not to mention out of. Evidence of this can still be seen in the old track out of the valley where jeep drivers had to fight their way up the slope.

Now there are two good routes and one for big jeeps only.

Because of the hydropower project at Kárahnjúkar the main road has been built up and paved as far as the rim of the valley, just to the north of the mountain Lambafell. Approaching from the north, you

There is plenty of room in the main pool in the stream by Laugarvellir. Near the pool is an old round-up hut sometimes used as a changing room. Photo: Þóra Sigurbjörnsdóttir

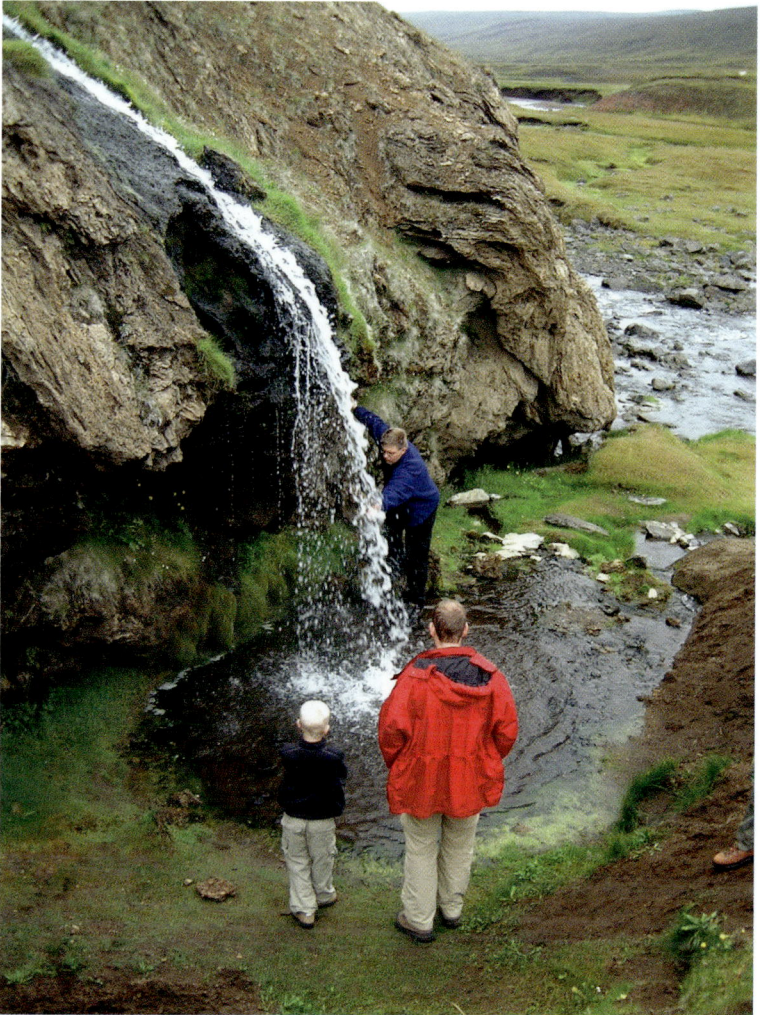

*Checking the temperature of the natural shower below Laugarvallalaug.
Photo: Jón G. Snæland*

take the new road south off the F910 at Fiskidalsháls ridge. South of Reykjará you can fork right onto a jeep track and make your way south along the Laugarvalladalur valley, but this track is just barely passable and would certainly not be passable by a jeep with a trailer tent or suchlike in tow.

Farming in Laugarvalladalur

In the year 1900, a young couple obtained a smallholding in Laugarvalladalur, from the farm Brú, and built their croft. It was a promising place, grassy and giving good hay. It is not known to have been farmed before that, although the grazing land was unbroken all the way south to Sauðárdalur valley and there was a hot spring right beside the house. It didn't snow all that much there and they could depend on a fair bit of winter grazing. They were hard-working and managed to increase their flock to begin with. The children came one after another and after six years they had two alive but had lost two in their infancy, as so often happened in those days. In early spring of 1906 there was a raging blizzard. The young farmer's sheep, along with some sheep he was feeding for other farmers, had been outside when the blizzard came and were driven away by the weather. The farmer went out in the storm to try to herd the sheep home again. He didn't manage to round them up and they were driven by the storm all the way to the river Jökulsá á Brú, where they all drowned. This was more than the farmer could take. He ate some fox poison, strychnine. His wife and mother-in law found him dying and tried to help him, but he soon met his death. The young widow now had to walk the 20 km to the farm Brú, leaving her children in the care of their grandmother. This was the end of farming in Laugarvalladalur. "But people often see him walking about here," said Sigvarður, the farmer at Brú, as serious as a grave, "especially if they spend the night here on their own."

www.nat.is

Near the Laugarvalladalur pool there is a dilapidated round-up hut, a primitive camping ground and one privy. The hut might be used for changing clothes, but most people prefer to lay their clothes on the grass. The pool is a dammed stream along the edge of the former homefield. It is 50 m long, 2-4 m wide and about 50 cm deep. The 70°C hot water comes up from the bottom but it mixes with the stream to give a water temperature of 37-38°C – perfect for bathing. The water gets quite cloudy, so the thing to do after bathing is walk down below the bank to where the water from the stream runs off a cliff. There you can rinse off the dirt in a natural warm shower.

The South

Seljavallalaug

GPS N63 34.013-W19 36.379
80–100 meters asl

The Seljavallalaug pool is below the eastern Eyjafjöll mountains, in a valley between Lambafellsheiði to the west and the mountain Raufar-fell to the east. Leave Highway 1 on road no. 242, branching left to pass the swimming pool and camping grounds, about 3 km, til the road ends below a group of summerhouses. From there walk along a pleasant path up along the Laugará river, a 15 to 20 minute walk. The pool is built against a cliff, concrete walls on three sides. The hot water seeps out of the cliff which forms one of the long walls of the pool. By the end of the pool is a concrete building with separate men's and women's changing rooms. The pool is 28 x 10 m. It was built in 1923 by the Eyjafjöll Youth Association, which still owns it, in a place where there was already a small turf and stone-built pool. In 1927 the local council was the first in Iceland to decide children would be required to

The left wall of the pool is a cliff from which warm water seeps into the pool. Photo: Jón G. Snæland

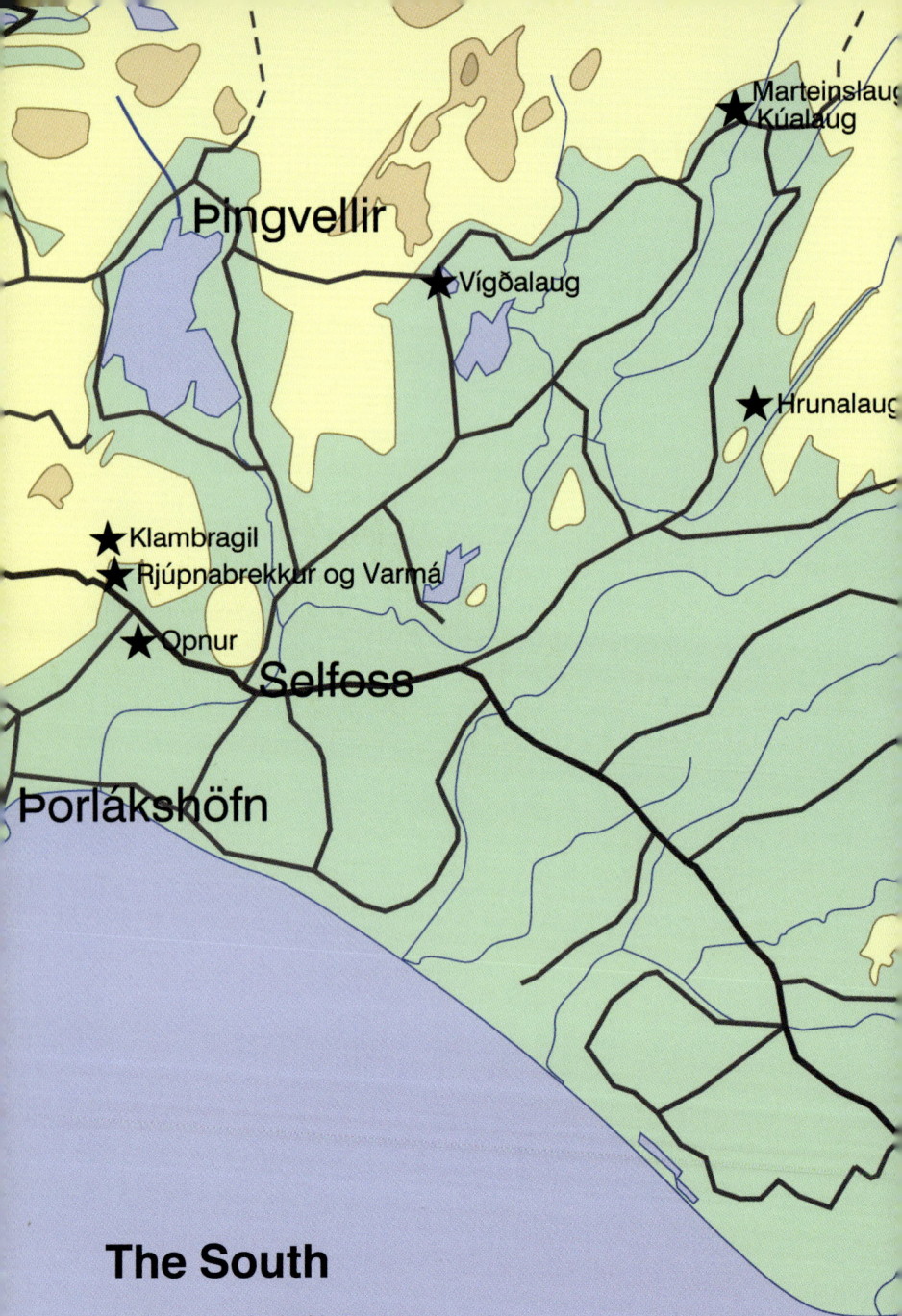

Marteinslaug
Kúalaug

Þingvellir

Vígðalaug

Hrunalaug

Klambragil
Rjúpnabrekkur og Varmá

Opnur

Selfoss

Þorlákshöfn

The South

Pools are marked with a star.
Pools marked with a green star
are not specifically covered.

★ Þjórsárdalslaug

★ Landmannalaugar

★ Strútslaug

Eyjafjallajökull

Mýrdalsjökull

★ Seljavallalaug

Vík

Hafsteinn, Ársæll, Aron and Arnór bathing in Seljavallalaug pool. Photo: Jón G. Snæland

learn to swim. The modern swimming pool by the campsite at Selja-vellir is fed from the same source, but in recent years neither that pool nor the campsite have been open.

Þjórsárdalslaug

GPS N64 09.646-W19 48.684 WGS84
185 meters asl

From Highway 1 take road no. 30, then 32 up along the west of the Þjórsá river, heading for Búrfell. Shortly before you reach Fossá take a left and drive 8.8 km to the end of the gravel, all-vehicle road, keeping right along the south of Reykholt ridge. The road ends at the pool. It was built in 1973 by the crew that built the Búrfell hydroelectric plant, and belongs to Landsvirkjun. The pool is 25 x 13 m, up to 1.8 m deep. There is also a hot tub, changing rooms and a shower. It has been kept open only in the summer.

Þjórsárdalslaug was built in 1973 by the crew that built the Búrfell hydro-electric plant. Photo: Jón G. Snæland

Hrunalaug

GPS N64 08.034-W20 15.428 WGS 84
138 meters asl

Hrunalaug is just a few kilometers beyond Flúðir. Drive along road no. 344 to the church farm Hruni, then over a cattle grid onto road no. 345. After 2-300 m there is a little parking area and from there a 5 minute walk to the pool. The pool is in a grassy dell where a little warm stream forms a long, narrow pool, stone-built, above a little old concrete building with a turf-clad roof. This pool is 4.5 x 1.45 m with a depth of about 50 cm, the water flowing up through the gravel bottom at 3.3 l/sec. The water then flows through the little building – which is now used as a changing room – and out into a concrete cistern, a stand-in pool a little over a meter in depth, 1.64 x 1.50 m in size. You walk straight out of the changing room and down a step into it, and

The old building works fine as a changing room from which you enter the standing pool in front of it. Photo: Jón G. Snæland

The larger pool is above the building, and it can be pleasant to let the wind dry you or the sun bake you in the beautiful surroundings. Photo: Jón G. Snæland

there is room for 2-4 people, standing, but the other pool can take 6-8. The temperature of the water is 37-38°C. There is a bench inside the building – which has in fact only three walls – where you can leave your clothes if you don't choose to simply lay them on the grass.

In former times this pool was probably used as a sheep bath.

Vígðalaug

GPS N64 12.938-W20 43.782 WGS 84

71 meters asl

Vígðalaug is at Laugarvatn on road no. 37. Drive into the village and turn down east of the building with all the dormers to the lake. The street ends there by a signboard above the boathouse. The pool is a few meters from the parking area. It is stonebuilt, circular and a warm stream runs through it. It is about 160 cm in diameter and 30 cm deep. Nearby are six large stones, partly buried by earth and grass. They are called "the corpse stones" and are said to have been used as a platform for the biers of the Hólar bishop, Jón Arason, and his sons Ari and Björn, who where all beheaded at Skálholt in 1550. Their bodies were washed with water from the pool after having lain in the ground over the whole winter.

This pool was also often used for christening people long ago. The water is considered to have supernatural healing powers like Krosslaug in Lundarreykjadalur, and was considered especially good for aching eyes.

The body of bishop Jón Arason was washed with the consecrated water from the pool after he was beheaded. Photo: Jón G. Snæland

Kúalaug

GPS N64 19.605-W20 16.924 WGS 84

113 meters asl

Kúalaug is in the Haukadalur valley above Geysir. You drive past Geysir and left along road no. F333, then along a gravel road towards the Haukadalskirkja church. There you will come to a reforestation area and a signboard with information about it. Drive into the wood

Kúalaug in Haukadalur. Photo: Jón G. Snæland

Kúalaug in winter. Photo: Jón G. Snæland

and past a memorial to the Dane, Kristian Kirk, who had owned the farmland of Haukadalur and given it to the Iceland Forest Service. Shortly thereafter you reach a dip where you park below the church. A few meters above the road on your left are two pools. Kúalaug is circular and stone built, just over a meter in diameter and 50-60 cm deep with a water temperature of 39-40°C. The other pool, nearby, is oblong, about 2 m long and 50-60 cm wide and a little deeper than Kúalaug. Both pools have earth bottoms and cloud up with use. Clothes can be laid on the grass.

Marteinslaug

GPS N64 19.634-W20 16.769

120 meters asl

Drive past Geysir and take road no. F333 on the left, then along a gravel road towards the Haukadalskirkja church. There you will come to a reforestation area and a signboard with information about the wooded area.

This is a good place to leave your car and study the signboard because you are going to walk along a pleasant path through the wood to reach the pool, which is on the bank of the river Kaldilækur. The naturalist and doctor, Eggert Ólafsson and Bjarni Pálsson, who thoroughly investigated the whole country in the mid 18th century and wrote of what they found, say that the people had great faith in Marteinslaug as a healing pool. It was thought to be named either for a 16th century bishop in Skálholt named Marteinn or for the much earlier (ca

Decks and grids cover Marteinslaug for some unknown reason. Photo: Jón G. Snæland

Anna Soffía standing on the lid of Marteinslaug.. Photo: Jón G. Snæland

4th century) St. Martein of Tours, who was uncommonly popular in Iceland well into the 18th century.

Marteinslaug has often changed over the years and there are now concrete cisterns which are closed with wooden grids. The pool is now a protected conservation site under the auspices of the Iceland Forest Service.

Opnur

GPS N63 58.875-W21 10.591 WGS 84
17 meters asl

Driving east on Highway 1, pass Hveragerði, cross the river Varmá and take a right onto the road to the farm Vellir. Drive past a horse ranch and a red farmhouse onto a newish road, about 600 m altogether. In a bend in the road below the red house you can leave your vehicle. Two fences meet there, and it's an ideal place to climb over. Then you walk about 300 m across pretty wet tussocked ground, crossing two ditches, so it's a good idea to have appropriate footwear. The old pool at Opnur is about 32 x 8m and up to 50 cm deep. A hot stream, about 28-30°C, runs into it on the north side. The pool gets cloudy when used. There were swimming lessons there in the early 20th century, but teaching stopped several decades ago. However, you can see that the pool was to an extent man-made, since the banks are still pretty straight.

Opnur is an exceptionally large pool and must have been considered huge around 1900 when it was used for teaching. Photo: Jón G. Snæland

Klambragil

GPS N64 02.915-W21 13.352

272 meters asl

Klambragil is below the south side of Ölkelduháls ridge, usually reached by one of two routes.

You can drive through the town of Hveragerði, beyond the town and towards the horse stables area. Where the road forks, carry on along the Varmá river until the road ends in a parking lot on the river bank. From there it is about a 3.3 km walk to Klambragil, along a former jeep track up the Rjúpnabrekkur slopes and in along the mountain Dalafell.

It is also possible to turn off Highway 1 up on the Hellisheiði heath, taking the road built by Reykjavik Energy up onto Ölkelduháls. When you reach Bitra, near Molddalahnúkar you can drive up a short built-up road to park. From the main road it is about 3.2 km to Klambragil. There are signs and markings along the way. The path down into Klambragil is good but pretty steep, taking you about 50-60 m lower

At the inner end of Klambragil there are a lot of steam and mud hot springs. Photo: Jón G. Snæland

One of the pools in the hot stream in Klambragil. Photo: Jón G. Snæland

down. Approaching this way you first come across a lot of steam and mud hot springs at the inner end of the gorge, which are well worth examining. There are paths both along the edge of the gorge and along the stream, but the latter is pretty wet. There are both hot and cold water springs in several places along the stream so the water temperature varies a lot. When you get further out the gorge, one hot stream joins another stream that runs in from Dalaskarðshnúkur. Just before they join up the hot stream has been dammed with rock walls in two places. The water is about 40 cm deep and the water temperature is a comfortable 30-35°C. There are in fact several dams in the stream. Above the pool to the south-east is Reykjavik Energy's lodge, in Dala-skarð pass. It is about 500 m away but difficult to see as it blends with the landscape. The lodge is called Dalasel (GPS N64 03.116-W21 12.999) and can sleep 12 in bunks. It is open to walkers.

Rjúpnabrekkur and Varmá

GPS N64 01.509-W21 12.685 WGS 84

89 meters asl

Drive through the town of Hveragerði, and up along the river Varmá on the street Breiðamörk. Then drive towards the horse stables area, fork left and drive to the parking lot by the river Hengladalsá. There you cross the river on a good strong foot bridge and head for the Rjúpnabrekkur slopes along the path leading to a borehole and a platform. On the way you pass some signboards and below them is an old jeep track leading to Rjúpnabrekkur and the Reykjadalur valley, where there is also a hot stream. Streams running from a number of hot springs just above the platform join together below the slope. There are also warm springs below the platform and all these geothermal pheno-

The stream at the roots of Rjúpnabrekkur runs into the river Varmá, and occasionally gets blocked on its way there. Photo: Jón G. Snæland

There are many places at the roots of Rjúpnabrekkur where hot water comes out of the ground. Photo: Jón G. Snæland

mena create a warm stream at the bottom of the hill. The stream does occasionally get blocked, but spring spate usually clears it. The stream is quite deep, but the nearer it gets to the river Hengladalsá the cooler it gets. A little pool near the upper end has been recorded at 33°C.

Below Rjúpnabrekkur, the river Hengladalsá, a stream out of Djúpagil gorge, and the river Grænadalsá combine to form the river Varmá, which flows through the town of Hveragerði. These and many other hot springs and warm streams run into the Varmá, which also has hot wellsprings in the river itself. The water in the river can thus be a good temperature for swimming when it is not too deep, being then around 20°C. The most popular swimming place is by the waterfall Reykjafoss, where the brave jump down from the top of the falls, but there are also many good places above the falls.

Reykjanes

Skátalaug

GPS N63 54.235-W22 02.603 WGS 84
160 meters asl

Skátalaug is to the south of the lake Kleifarvatn. Taking road 41 out of the greater Reykjavík area, turn onto road no. 42, the Krísuvík road. Drive all the way south of the lake and south of the large hillock Bleikhóll (Pink or Pale Hillock) on your right. There you turn west, taking a right off the highway towards the mountain side. The track begins at N63 54.281-W22 02.348. Drive a short way and turn left at N63 54.274-W22 02.595. The pool itself is at N63 54.235-W22 02.603. The distance is 286 m, and not passable by sedan cars except in the best of weather. The pool is not easy to see from the end of the track, so walk towards the highway from the parking place.

The pool is fed 34°C hot water through 4 iron pipes at a rate of 1.5 l/sec. The water runs into a small (1.5 x 1.5, 50-60 cm deep) basin at the end of the pool. It is partly separated from the main pool by a low rock wall. The main pool is 13-14 m long, 5 m wide and about a meter

It can be hard to find Skátalaug without GPS. Photo: Jón G. Snæland

The ground around the pool is pretty marshy. Photo: Þóra Sigurbjörnsdóttir

in depth, with room for several dozen bathers at once. There is a great deal of algae on the surface of the main pool. The banks are grassy but mostly marshy. The bottom of the pool is dirt and rock and the pool gets very cloudy.

The Blue Lagoon

GPS N63 52.797-W22 26.940

37 meters asl

Not long after the geothermal power station was built at Svartsengi around 1980, warm seawater started streaming up to the surface. This subterranean seawater is part of an ecosystem that came into being through the interplay of science and nature. The water comes from a depth of 2000 m where it is heated by natural forces. The temperature at this depth is 240°C and the pressure 36 times greater than on the earth's surface. When seawater comes into contact with the cooling lava it absorbs various minerals, creating this unique natural spring known for active ingredients and curative powers. As it rises it combines with surface water. The mineral composition of the final product is unique, with a very high percentage of silica and 2.5% salt ratio, about one third the amount of salt found in sea water.

It wasn't long before people suffering from psoriasis or other skin diseases discovered the curative effects of this blue water. By 1982 they

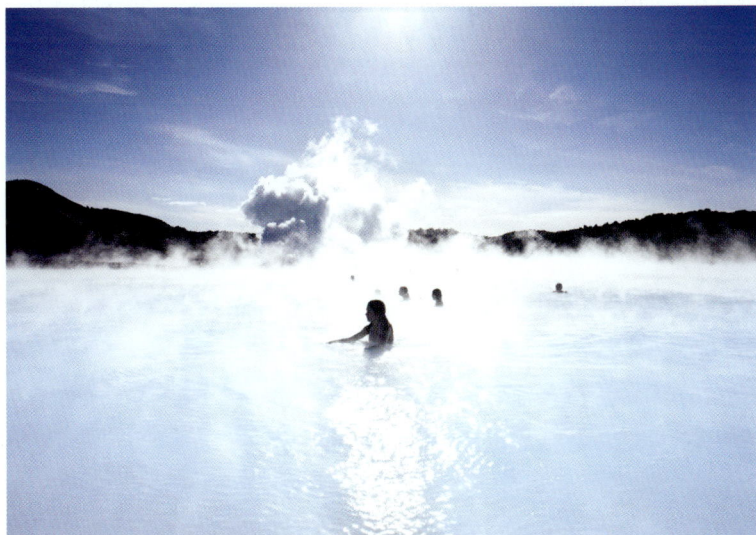

In the Blue Lagoon. Photo: Bláa Lónið

In the Blue Lagoon. Photo: Bláa Lónið

set up a building beside this lagoon that came from the power station. In 1992 a company was formed, Bláa Lónið ("The Blue Lagoon") to run the place. The operation has grown steadily, and in 1999 they opened a new service centre with much improved bathing facilities, which have since been extended. Now there are restaurants, convention facilites and first class services, and Bláa Lónið has become one of Iceland's greatest tourist attractions.

Nauthólsvík

GPS N64 07.276-W21 55.676
On the shore

In the year 2000 the city of Reykjavík opened the so-called warm beach at Nauthólsvík inlet, south and east of the airport. They built two great breakwaters out beyond the inlet, leaving a gap so there is still high and low tide in the big lagoon they created. The shore of the lagoon was filled with golden sand and hot water pumped into the sea there, so it reaches temperatures up to 18-20°C. Thus they created a lovely beach of a type more often found in southern climes, which is now so popular that it gets over 100.000 visitors annually.

In 2001 a sevice centre providing showers, changing rooms and a snack bar was opened. Besides the lagoon itself there are two large hot tubs, one (38°C) by the service centre, the other (25°C) on the foreshore. The actual beach is open all year round, but hot water only runs into it in the summer, May 15th to August 31st. In the winter the

Two great breakwaters were built to define the warm beach lagoon. Photo: Jón G. Snæland

The upper hot tub by the service centre. Here it is being used by the people who regularly swim in the lagoon in the winter. Photo: Sif Friðleifsdóttir

hardy sea bathers make use of Nauthólsvík. It is open weekdays from 10.00 to 20.00, weekends to 18.00, but to 20.00 when the weather is really fine.

Kvika

GPS N64 09.743-W22 00.496 WGS 84
On the foreshore

The pool Kvika, on the Seltjarnarnes peninsula at the western end of Reykjavík, is probably the smallest thermal pool in the country. It is 80-90 cm in diameter and 25-30 cm deep. Drive out along the northern shore on Norðurströnd street, most of the way to the lighthouse in Grótta. Once past the residential area you see a white concrete pump house (borehole) belonging to the Seltjarnarnes Heating Utility. There you can park and walk to an old shark-drying rack on the foreshore. Below the rack is the pool. It is an artwork by the visual artist Ólöf Nordal, intended for bathing feet. It has been hewn out of a huge dolerite boulder and with a little imagination can be seen to resemble an enormous wooden ladle. A dim electric light from under the water illuminates the steam. This artwork was unveiled in June 2005 on Reykjavík Culture Night.

Siv Friðleifsdóttir warming her feet in Kvika. Photo: unknown

Common endings

-**alda** swell
-**á** river
-**dalur** valley
-**ey** island
-**fell** mountain
-**fjall** mountain
-**fjörður** fjord
-**fljót** river
-**foss** waterfall
-**gil** gorge
-**gjá** gorge
-**heiði** heath
-**hellir** cave
-**hver** hot spring
-**jökull** glacier

-**kofi** hut
-**kvísl** branch
-**laug** pool
-**lækur** creek
-**lón** lagoon
-**strönd** shore
-**brekkur** slope
-**vað** ford
-**leið** road, track
-**skáli** lodge
-**skarð** pass
-**sýsla** county
-**vatn** lake
-**vegur** road, route
-**vellir** field

Index